Carpentry and Joinery: A Multi-Questions Course

F. A. Inott, FBICC, ARSH
Formerly Senior Lecturer, Timber Division,
Shirecliffe College of Further Education, Sheffield

Carpentry and Joinery: A Multi-Questions Course

GRANADA
London Toronto Sydney New York

Granada Publishing Limited – Technical Books Division
Frogmore, St Albans, Herts AL2 2NF
and
36 Golden Square, London W1R 4AH
866 United Nations Plaza, New York, NY 10017, USA
117 York Street, Sydney, NSW 2000, Australia
100 Skyway Avenue, Rexdale, Ontario, Canada M9W 3A6
61 Beach Road, Auckland, New Zealand

ISBN 0 246 11929 2

First published in Great Britain 1974 by Crosby Lockwood Staples
(original ISBN 0 258 96952 0)
Re-issued 1982 by Granada Publishing

Printed in Great Britain by
Richard Clay (The Chaucer Press) Ltd,
Bungay, Suffolk

Contents

Preface

This book is designed to assist students of carpentry and joinery to assess their progress in technical studies, so that specific weaknesses can be quickly identified for further study.
The standards of the 500 questions in this book have been graded, in numerical order, to meet the students' requirements to craft certificate level.
An index of questions related to particular topics has been included to enable a student to select specific questions on particular subjects.
Students are advised to use a separate notebook to record the question number and their selected answer, i.e. A, B, C or D (one answer only), then check the results under the 'Key Answers'. The topics of incorrect answers should be noted and revision study carried out before re-attempting the questions.
Teachers may find this book useful as a basis for developing their own programme of progress tests.

F. A. Inott

Index to Question Topics by Question Numbers

Safety and Health

1 2 41 62 64 181 184 185
186 187 189 196 197 198 294 299
301 306 312 326 340 344 400 403
448 36 50 51 56 60 151 152 169
180 207 208 245 261 275 278 281
305 314 348 360 381 385 399 424
464 491

Hand Tools

5 6 10 12 18 102 108 134
183 202 230 260 265 290 304 387
443 452 488

Power Tools

17 34 127 181 302 307 313 331
341 345 355 357 359 404 459

Calculations

4 5 8 13 21 24 26 29
75 92 94 95 100 103 104 125

Wood Technology

15 19 23 30 32 43 73 91
98 191 192 210 216 222 224 227
293 351 429 458 500

Materials

35 45 121 192 193 194 195 210
212 218 220 225 226 228 239 291
298 310 325 393 409 430 445 447
449 458 476 479 484

First and Second Fixings

3	11	16	27	33	109	122	123
129	132	133	134	135	136	139	143
235	240	247	250	270	329	335	394
84	85	86	89	90	114	322	324
325	337	350	413	414	415	418	449

Adhesives

211	213	215	217	219	221	223	225
228	231	271	282	300	390	398	403

Doors: Frames and Linings

96	135	138	143	209	232	233	234
236	241	242	243	244	245	250	252
253	254	255	258	269	271	354	361
362	363	364	365	372	378	391	461
463	466	467	469	470	485		

Floors

63	76	77	78	80	81	82	83

Roofs

46	47	48	49	52	53	54	55
57	59	61	66	67	68	69	70
71	72	316	317	319	320	327	328
329	330	335	342	411	416	427	435
437	438	439	440	445			

Windows: Wood Trim

214	229	237	246	247	248	249	251
256	367	368	369	370	375	376	379
386	388	395	396	402	474		

Ironmongery

88	126	132	138	141	224	249	267
328	342	350	394	395	401	402	441
461	463	464	468	470	471	473	

Partitions: Panelling

110 115 116 117 394 460 481 482
483 487

Gates and Fencing

113 118 119 120 383

Temporary Timbering: Formwork

144 145 146 147 148 149 150 338
339 344 349 426 442 444 446 448
492 494

Cupboards: Fitments

256 257 258 259 260 266 267 268
270 279 280 282 283 284 285 290
291 298 373 377 382 384 389

Stairs

392 400 483 489 490 493 495 496
497 498 499

Timber Framed Buildings

297 332 432 433 434 437 479 484

Questions 1–500

1 A scaffold must be fitted with toeboards and handrails when it reaches a height of

- *A* *$1\frac{1}{2}$ metres*
- *B* *2 metres*
- *C* *3 metres*
- *D* *1 metre*

2 The correct angle for using a ladder when it is placed against a wall is

- *A* *7 up, 2 out*
- *B* *6 up, 1 out*
- *C* *4 up, 1 out*
- *D* *5 up, 3 out*

3 Which one of the following activities would be classed as carpenter's work?

- *A* *jointing narrow boards for a wide shelf*
- *B* *making a larder casement window*
- *C* *mortising stiles for a panelled door*
- *D* *levelling and fixing floor joists*

4 The SI unit of length is

- *A* *millimetre*
- *B* *metre*
- *C* *litre*
- *D* *kilometre*

5 Which of the following planes has one iron only?

- *A* *block*
- *B* *badger*
- *C* *smooth*
- *D* *jack*

6 A carpenter's plugging chisel has developed a 'mushroom' head. Before using further you should

- *A* *re-temper the cutting edge*
- *B* *re-grind the cutting edge*
- *C* *straighten the shank*
- *D* *re-dress the head*

7 An isosceles triangle can be defined as having

- *A* *one angle a right angle and all sides of different length*
- *B* *all angles different and all sides different*
- *C* *two angles equal and two sides equal*
- *D* *all angles equal and all sides equal*

8 A corridor floor is 2·200 m wide by 11·000 m long, area will be

- *A* *22·100* m^2
- *B* *24·200* m^2
- *C* *19·750* m^2
- *D* *33·300* m^2

9 Which of the following geometrical planes has four sides?

- *A* *triangle*
- *B* *segment*
- *C* *quadrilateral*
- *D* *polygon*

10 The grinding angle for the cutter of a smoothing plane is

- *A* *25°–30°*
- *B* *50°–55°*
- *C* *35°–40°*
- *D* *30°–35°*

11 A carpenter's work of cutting and fixing roof timbers is known as

- *A* *rough fixing*
- *B* *first fixing*
- *C* *second fixing*
- *D* *roof capping*

12 A handsaw with 3·5 teeth per 25 mm would be suitable for

- *A* *ripping timber*
- *B* *cross-cutting timber*
- *C* *cutting plywood sheets*
- *D* *cutting dovetails*

13 A 3 m measuring tape is approximately equivalent to

- *A* *7·5 ft*
- *B* *9 ft*
- *C* *10 ft*
- *D* *12 ft*

14 Which of the following quadrilaterals has four equal sides?

- *A* *rhomboid*
- *B* *rectangle*
- *C* *trapezium*
- *D* *rhombus*

15 The actual growing part of a living tree is known as

- *A* *heartwood*
- *B* *cambium layer*
- *C* *medullary ray*
- *D* *bark and bast*

16 A splay scarf joint is used for lengthening

- *A* *a roof ridge*
- *B* *a wall plate*
- *C* *a floor board*
- *D* *a door jamb*

17 A 13-amp square pin plug is to be fitted to a cable with wires to the new code. Which colour wire must be connected to the earth pin?

- *A* *brown*
- *B* *blue*
- *C* *grey*
- *D* *yellow and green*

18 When grinding tools, an essential safety precaution is to wear

- *A* *protective helmet*
- *B* *protective goggles*
- *C* *waterproof gloves*
- *D* *steel-tipped boots*

19 Western red cedar timber is

A *a dense hardwood*
B *an unobtainable timber*
C *a rot-resisting timber*
D *suitable for floor boards*

20 The SI metric unit for measuring mass is

A *kilometre*
B *gramme*
C *litre*
D *kilogramme*

21 A cylindrical storage bin has a circumference of 6·284 m and axis height of 3·000 m. The area of the curved surface, not including the bottom, will be

A $18{\cdot}852\ m^2$
B $2{\cdot}142\ m^2$
C $20{\cdot}450\ m^2$
D $9{\cdot}050\ m^3$

22 Softwood is a general classification of timber obtained from trees with

A *broad-shaped leaves*
B *rough bark*
C *needle-like leaves*
D *light-coloured wood*

23 Two common rafters meet at the ridge of a roof to form an angle of 120°, as in Figure 1. This angle is called

A *an acute angle*
B *an obtuse angle*
C *a scalene angle*
D *an equilateral angle*

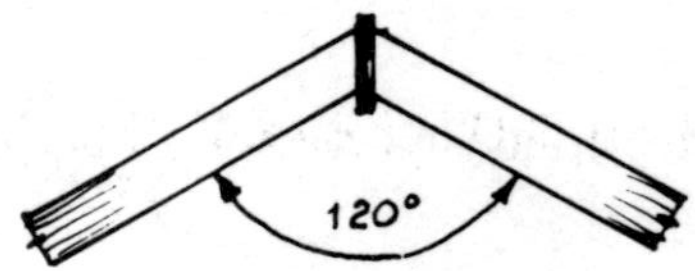

Figure 1

24 In SI units a metre consists of

- *A* *1000 mm*
- *B* *1000 cm*
- *C* *1000 cc*
- *D* *500 mm*

25 A table scarf joint used for lengthening structural timbers is designed to resist

- *A* *tension*
- *B* *shear*
- *C* *splitting*
- *D* *compression*

26 $\pi = 3 \cdot 142$; this is derived by

- *A* *multiplying the circumference of a circle by the diameter*
- *B* *dividing the circumference of a circle by the radius*
- *C* *dividing the circumference of a circle by the diameter*
- *D* *multiplying the circumference of a circle by the radius*

27 A tusk tenon joint is used in construction of

- *A* *roof trusses*
- *B* *doors*
- *C* *windows*
- *D* *floors*

28 Which of the following triangles, whose sides are given, is a right angled triangle?

- *A* *4 : 5 : 6 m*
- *B* *6 : 8 : 10 m*
- *C* *7 : 11 : 13 m*
- *D* *6 : 9 : 11 m*

29 Five-eighths expressed as a percentage, is

— *A* 62·5%
B 66%
C 75%
D $33\frac{1}{3}$%

30 Figure 2 illustrates

A laminboard
B particle board
C batten board
— *D* blockboard

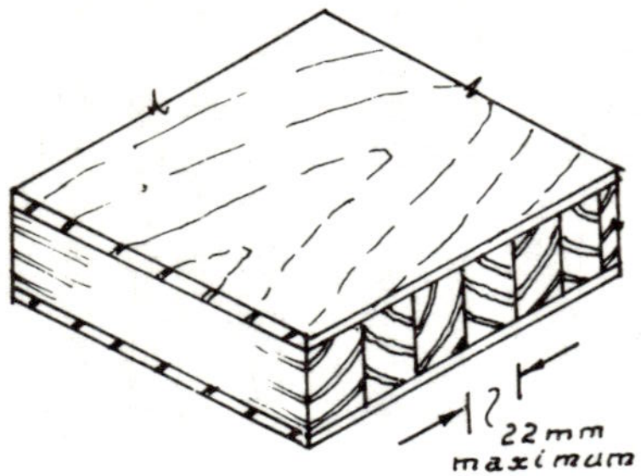

Figure 2

31 Which machine does a sawyer use for breaking down a softwood log into 'through and through' sawn boards in one operation?

A log band saw
— *B* log frame saw
C circular saw
D horizontal reciprocating saw

32 Which method of conversion would produce a board as shown in Figure 3?

— *A* rift sawn
B slash sawn
C plain sawn
D flat sawn

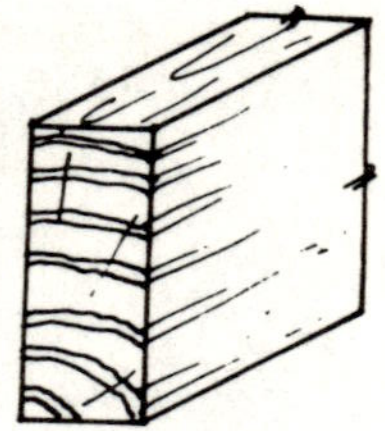
Figure 3

33 Which manufactured board should be damped before fixing?

A low density insulation board
B veneered chipboard
C formwork plywood
— *D* standard grade hardboard

34 The correct voltage for power tools on a building site is

A *20 V*
B *100 V*
C *110 V*
D *240 V*

35 The 'nominal size' of timber means

A *sawn size of timber section*
B *BS standard size*
C *finished size of timber section*
D *approximate size of timber section*

36 The nett area of the triangular gable shown in Figure 4 is

A $38{\cdot}500\ m^2$
B $21\ m^2$
C $42\ m^2$
D $22{\cdot}750\ m^2$

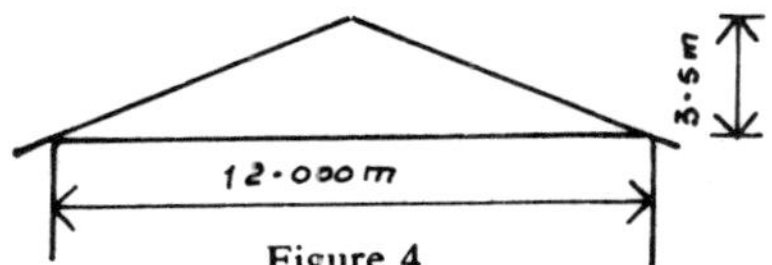

Figure 4

37 The best method of applying timber preservative to obtain the deepest penetration is by

A *spraying*
B *dipping*
C *brushing*
D *pressure impregnation*

38 Tar oils and water-borne salts are types of

A *preservative for timber*
B *sap in trees*
C *timber fertiliser*
D *priming paint*

39 Surface checking on timber is

A *fungal decay*
B *separation of the growth rings*
C *seasoning defect*
D *examining the timber surface for knots*

40 Figure 5 shows a timber with a waney edge defect. This is caused by

- *A* *over-economical conversion of the log*
- *B* *excessive sloping grain*
- *C* *rapid drying of the timber*
- *D* *careless sawing of the log*

Figure 5

41 Which is the correct way to lift a load, by hand, from floor level?

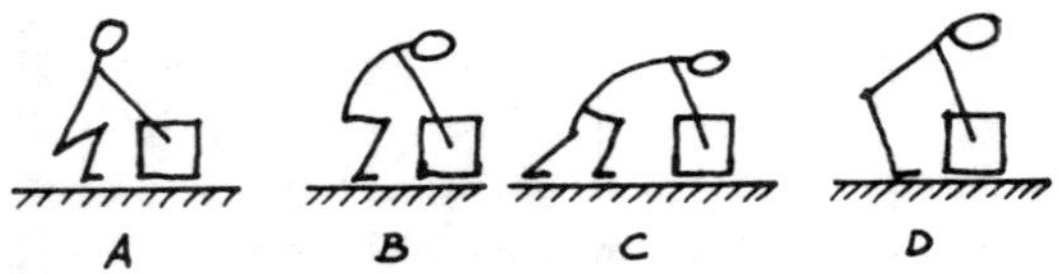

Figure 6

42 Figure 7 shows an unstable rectangular frame which requires a fifth member; this should be placed between

- *A* *g and c or a and e*
- *B* *h and d*
- *C* *b and f*
- *D* *b and f also h and d*

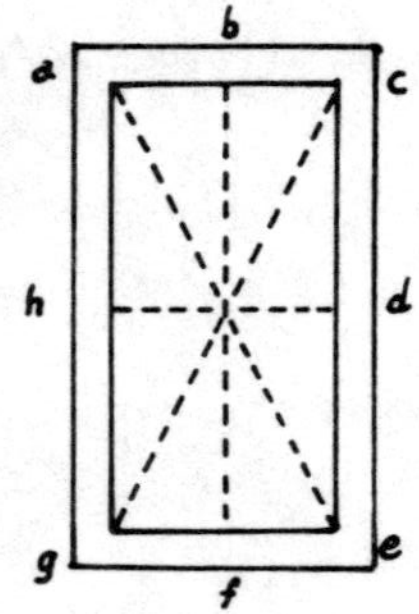

Figure 7

43 The best way of distinguishing between specimens of hardwoods and softwoods is by examining their

- *A* *colour*
- *B* *cellular structure*
- *C* *sapwood*
- *D* *density*

44 Which of the beam sections shown in Figure 8 would be the stiffest in resisting a bending load?

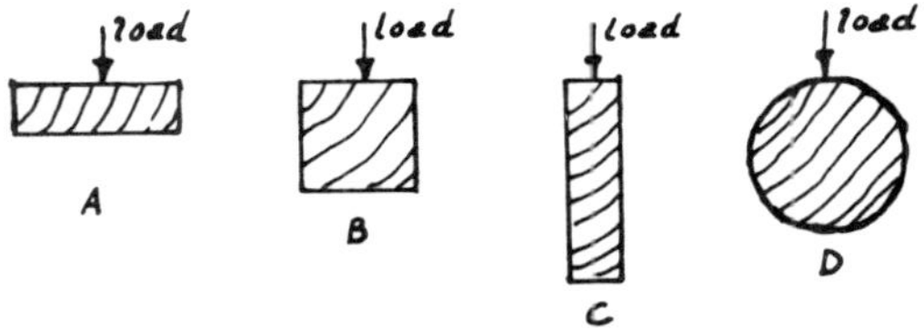

Figure 8

45 A timber merchant's size for sawn softwood boarding is

- *A* *over 100 mm wide, less than 50 mm thick*
- *B* *over 150 mm wide, over 50 mm thick*
- *C* *over 225 mm wide, 50 to 150 mm thick*
- *D* *50 to 75 mm wide, 50 to 75 mm thick*

46 Figure 9 shows the section of a single roof which is known as a

- *A* *collar roof*
- *B* *couple close roof*
- *C* *close roof*
- *D* *pent roof*

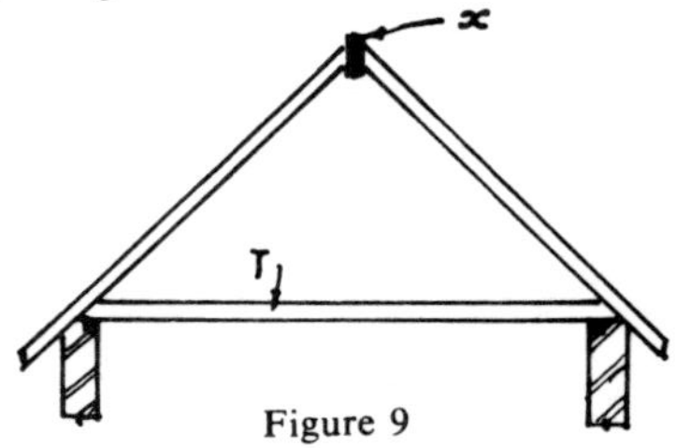

Figure 9

47 The member marked x in Figure 9 is termed

- *A* *ridge board*
- *B* *apex board*
- *C* *top board*
- *D* *purlin*

48 The foot of a common rafter is fixed to
- *A* *the brick wall*
- *B* *ceiling joist*
- *C* *fascia board*
- *D* *wall plate*

49 The member T in Figure 9 is used to support the ceiling, but its primary function is to
- *A* *reduce the length of the rafters*
- *B* *spread the weight of the roof*
- *C* *prevent the rafters from spreading*
- *D* *prevent the walls from spreading*

50 Which of the following formulae is used for calculating the area of a circle?
- *A* πD
- *B* πr^2
- *C* $2\pi r$
- *D* $\frac{\pi r}{2}$

51 The square root $\sqrt{\frac{16}{49}}$ is
- *A* $0{\cdot}75$
- *B* $\frac{9}{4}$
- *C* $\frac{4}{7}$
- *D* $\frac{7}{4}$

52 Which of the following handsaws is best for sawing roofing timbers to length?
- *A* *676 mm crosscut with 6 teeth per 25 mm*
- *B* *676 mm rip with 3·5 teeth per 25 mm*
- *C* *550 mm panel with 10 teeth per 25 mm*
- *D* *250 mm backsaw with 14 teeth per 25 mm*

53 ‘Barge boards’ is the name given to the boards
- *A* *used to fill-in the space under a stair*
- *B* *used to cover a roof before tiling*
- *C* *used to finish the verge of a gable roof*
- *D* *laid on a temporary working platform*

54 The angle of intersection between an inclined common rafter and the vertical ridge board will give the
- *A* *backing angle*
- *B* *plumb cut*
- *C* *purlin cut*
- *D* *valley angle*

55 Softwood is used in roofwork because
- *A* *it resists decay from fungi*
- *B* *it is more pleasant to work than hardwood*
- *C* *it has greater strength than hardwood*
- *D* *it is more economical than hardwood*

56 Figure 10 shows, in line diagram, a lean-to roof. The nett surface area of the roof will be
- *A* $18\frac{1}{3}\ m^2$
- *B* $32{\cdot}500\ m^2$
- *C* $16{\cdot}250\ m^2$
- *D* $9\frac{1}{6}\ m^2$

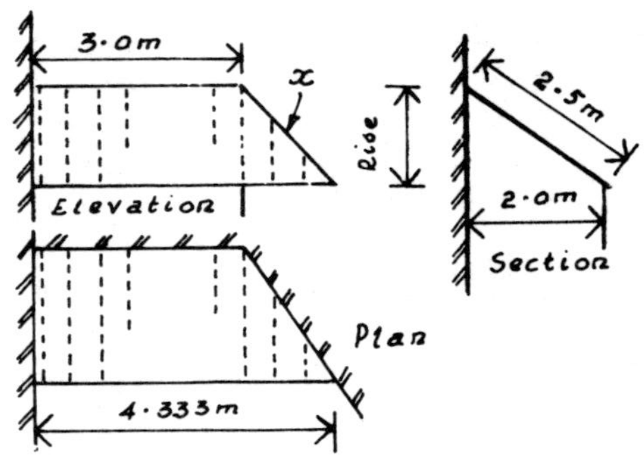

Figure 10

57 If the common rafters in Figure 10 are spaced at 433 mm centres, how many rafters, including the shortened ones, will be required?

A *14*
B *8*
C *12*
D *10*

58 The development of true shape of the roof surface in Figure 10 will give a geometrical plane known as

A *a trapezium*
B *a rhomboid*
C *a polygon*
D *a rectangle*

59 In Figure 10 the member spiked to the wall at x, to which the ends of the shortened rafters are fixed, is known as

A *barge board*
B *verge plate*
C *hip rafter*
D *valley rafter*

60 The rise of the roof in Figure 10 is

A *4 m*
B *1 m*
C *1·500 m*
D *0·500 m*

61 For a roof of 7·000 m span with a pitch of one third, the rise would be

A *2·333 m*
B *3·000 m*
C *4·750 m*
D *2·666 m*

62 Ladders used for access to a scaffold working platform must be tied at their top end to the scaffold, and extend beyond the platform for a minimum distance of

- *A* *1·066 m*
- *B* *750 mm*
- *C* *900 mm*
- *D* *4 rungs*

63 In hollow ground timber floors, there should be a clearance between the site concrete and the bottom edge of the joist of at least

- *A* *50 mm*
- *B* *150 mm*
- *C* *112 mm*
- *D* *125 mm*

64 Scaffold standard poles, when used on soft ground, must be supported by

- *A* *a reinforced concrete pad*
- *B* *a base plate resting directly on the ground*
- *C* *a base plate resting on a timber board 1 m long*
- *D* *a number of loose bricks laid side by side*

65 The triangle in Figure 11 is called

- *A* *an equilateral*
- *B* *a right angle scalene*
- *C* *an acute angle scalene*
- *D* *a right angle isosceles*

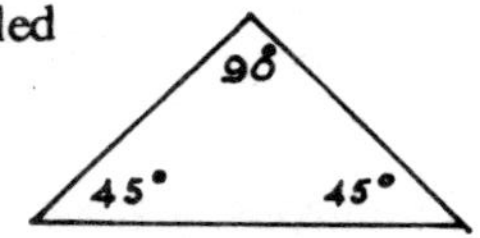

Figure 11

66 Figure 12 shows a typical section through the eaves of a 275 mm brick wall. This type of eaves is known as

- *A* *flush*
- *B* *closed projecting*
- *C* *overhanging*
- *D* *open*

67 In Figure 12 the joint J at the foot of the common rafter is known as

—*A* *birdsmouthed*
B *notched*
C *splay butt*
D *lipped*

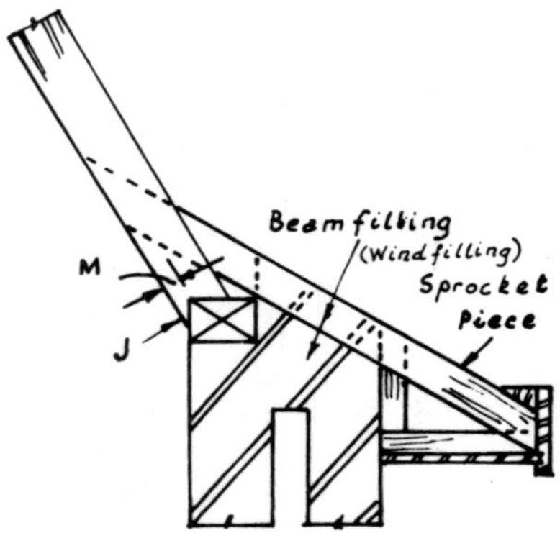

Figure 12

68 In Figure 12 the dimension M should have a proportion to the rafter depth of

A $\frac{1}{2}$
B $\frac{1}{4}$
—*C* $\frac{1}{3}$
D $\frac{5}{8}$

69 The beam filling shown in Figure 12 is intended to

A *finish the top of the wall*
B *support the roofing tiles*
C *hold the rafters in position*
—*D* *reduce the fire risks*

70 The sprocket piece shown in Figure 12 serves the purpose of

- A *lengthening short rafters*
- B *reducing the pitch of a steep roof*
- C *providing a wide soffit*
- D *a fixing for special roof coverings*

71 One purpose of a roof wall plate is to

- A *finish the top of a wall*
- B *distribute the load*
- C *support the purlin*
- D *hold down a D.P.C.*

72 A suitable joint for lengthening a roof wall plate would be

- A *lap halving*
- B *dovetailed halving*
- C *slot mortise and tenon*
- D *butt*

73 Which of the following timbers is most commonly used for carpentry work in house building?

- A *Yellow pine*
- B *Pitch pine*
- C *European redwood*
- D *Sitka spruce*

74 The angle of plumb cut for the common rafter in Figure 13 is

- A *80°*
- B *60°*
- C *30°*
- D *45°*

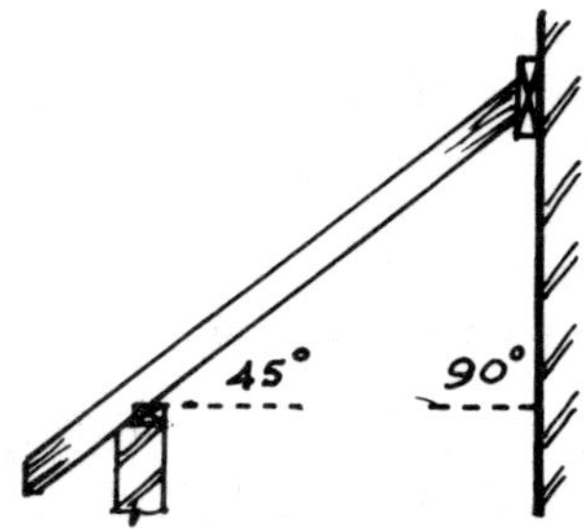

Figure 13

75 The nett area of roof boarding required for the roof surface in Figure 14 would be

- *A* *72 m^2*
- *B* *120 m^2*
- *C* *96 m^2*
- *D* *108 m^2*

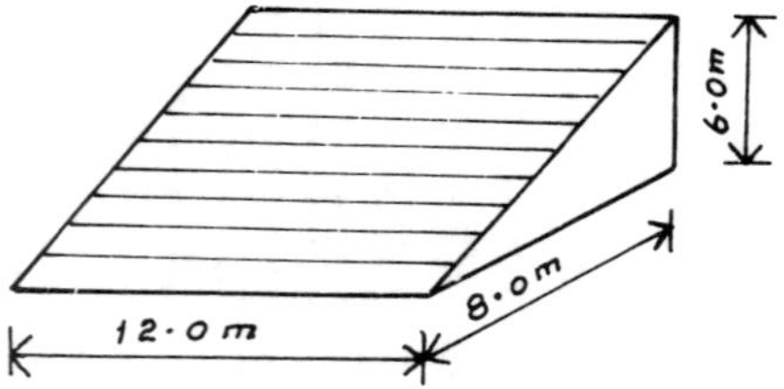

Figure 14

76 A suitable length for floor brads (nails) is

- *A* *1½ times thickness of the floor board*
- *B* *2 times thickness of the floor board*
- *C* *2½ times thickness of the floor board*
- *D* *3 times thickness of the floor board*

77 The purpose of a honeycombed sleeper wall is to

- *A* *provide intermediate support to a timber ground floor*
- *B* *support a non-loadbearing partition*
- *C* *provide a screen wall in a garden*
- *D* *support a fireplace hearth*

78 Figure 15 shows the section of a softwood floorboard with the tongues and grooves positioned off-centre. The reason for this is

- *A* *to permit easier laying of the boards*
- *B* *to provide sufficient wearing surface before exposing the tongue*
- *C* *because it is a custom of the trade*
- *D* *because the machine cutters cannot work in the centre of the board*

Figure 15

79 In Figure 15 the growth rings run from top to bottom faces; this

A *ensures the highest density*
B *eliminates sapwood*
C *is easier for nailing*
D *ensures minimum shrinkage across the board width*

80 A typical section size for Figure 15 when used in house floors is

A *25 × 125 mm*
B *25 × 65 mm*
C *38 × 100 mm*
D *13 × 150 mm*

81 An empirical rule which gives the approximate depth of a timber floor joist is

$$\text{Depth of joists (centimetres)} = \frac{\text{Span (decimetres)} + 2}{2}$$

Therefore the depth of joists for a span of 5·000 m is

A *250 mm*
B *270 mm*
C *260 mm*
D *190 mm*

82 The timber joists to a ground floor have to be notched to receive a 20 mm diameter water-pipe. The notches should be cut

A *on the top edge near the supported end*
B *on the top edge in mid-span*
C *on the bottom edge near the supported end*
D *on the bottom edge in mid-span*

83 A horizontal damp proof course should have a minimum height above ground level of

A *300 mm*
B *250 mm*
C *200 mm*
D *150 mm*

84 Site concrete under a timber suspended ground floor is intended to

- *A* *prevent growth of vegetation*
- *B* *form a useful storage area*
- *C* *prevent rising damp*
- *D* *discourage vermin nesting*

85 Air bricks are built into external walls at horizontal damp-proof course level to

- *A* *ventilate the cavity*
- *B* *drain away unwanted water*
- *C* *ventilate under a suspended ground floor*
- *D* *improve the thermal insulation of the wall*

86 The minimum distance from the front of a fireplace opening to a timber joist is

- *A* *500 mm*
- *B* *600 mm*
- *C* *300 mm*
- *D* *550 mm*

87 Figure 16 shows a circular pipe passing through a floor. The true shape of the hole in the floor will be

- *A* *circular*
- *B* *elliptical*
- *C* *diamond*
- *D* *egg shaped*

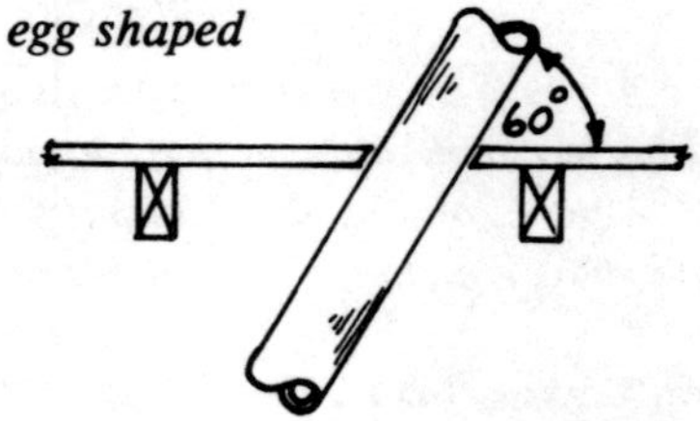

Figure 16

88 Which of the following nails should be used for fixing softwood tongued and grooved floorboards?

- *A* *round wire nails*
- *B* *galvanised clout nails*
- *C* *cut brads*
- *D* *oval wire nails*

89 A filler joist or saddle piece is used at the sides of a fireplace to provide a fixing for the floorboards. This should be set back from the fireplace opening at a minimum distance of

- *A* *300 mm*
- *B* *225 mm*
- *C* *75 mm*
- *D* *150 mm*

90 Softwood floor joists supporting ex 25 mm thick floorboards are normally spaced at centres of

- *A* *250 to 325 mm*
- *B* *350 to 590 mm*
- *C* *475 to 550 mm*
- *D* *575 to 650 mm*

91 Which of the following softwoods has the lowest density with the highest durability?

- *A* *Western red cedar*
- *B* *Pitch pine*
- *C* *European redwood*
- *D* *Columbian pine*

92 The logarithm characteristic for 293·45 is

- *A* *1*
- *B* *2*
- *C* *3*
- *D* *4*

93 A right angle is to be set-out on the ground using three pegs and a 12-metre tape. If the base of the triangle is 3 metres long, the ratio of the other sides will be

- *A* *3 : 1 : 8*
- *B* *3 : 2 : 7*
- *C* *3 : 3 : 6*
- *D* *3 : 4 : 5*

94 7^4 means

A 7×4
B $7 + 4$
C $7 \times 7 \times 7 \times 7$
D $7 \div 4$

95 $12\frac{1}{2}$% of £3·20 is

A *32p*
B *40p*
C *42p*
D *110p*

96 A rectangular door frame can be 'squared' by

A *using a 150 mm try square*
B *checking the diagonals are of equal length*
C *using winding laths*
D *checking the opposite sides are parallel*

97 An angle of 60° is to be set-out on a building site using timber stakes and tape measure. If two stakes are driven into the ground 5 metres apart, the measurements from each stake to a third stake will be

A *5 m and 5 m*
B *10 m and 10 m*
C *5 m and 12 m*
D *10 m and 15 m*

98 The stiffest carpentry timber is obtained from the

A *sapwood*
B *large branches*
C *crown of the trunk*
D *heartwood*

99 The theorum of Pythagoras relates to

A *angles of triangles*
B *parallelograms*
C *sides of a right angle triangle*
D *elliptical curves*

100 The square root $\sqrt{\frac{81}{121}}$ is

A $\frac{9}{11}$

B $\frac{8}{12}$

C $\frac{11}{9}$

D $\frac{9}{13}$

101 The three internal angles of an equilateral triangle are each

A *45°*
B *60°*
C *180°*
D *30°*

102 A wood rasp should always be used with

A *a flat face*
B *as many teeth as possible*
C *a properly secured handle*
D *a safe edge*

103 Using the extract in Figure 17 from a four-figure logarithm table, which is the logarithm of 42·63?

A *1·6291*
B *1·6297*
C *2·6291*
D *2·6398*

	0	1	2	3	4	5	6	7	8	9	1	2	3	4	5	6	7	8	9
41	6128	6138	6149	6160	6170	6180	6191	6201	6212	6222	1	2	3	4	5	6	7	8	9
42	6232	6243	6253	6263	6274	6284	6294	6304	6314	6325	1	2	3	4	5	6	7	8	9
43	6335	6345	6355	6365	6375	6385	6395	6405	6415	6425	1	2	3	4	5	6	7	8	9

Figure 17

104 The formula for calculating the volume of a right circular cylinder is

A $\pi r^2 h$
B $2\pi rh$
C πr^2
D $4\pi r^2$

105 Figure 18 shows the development of the sides for

A *an oblique rectangular prism*
B *a triangular prism*
C *a hexagonal prism cut by an inclined plane*
D *a right square prism cut by an inclined plane*

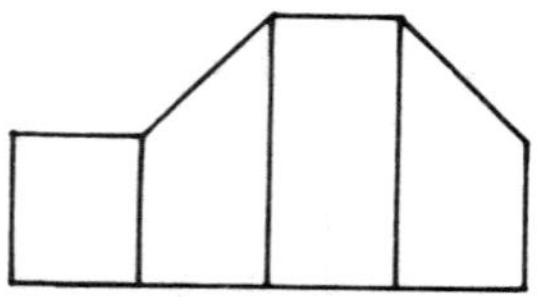

Figure 18

106 Braces in a garden gate are used to stop the gate from

A *dropping out of square*
B *warping*
C *winding*
D *being damaged by the weather*

107 The most satisfactory method of applying wood preservative to a timber garden fence is

A *spraying*
B *dipping*
C *pressure impregnation*
D *brushing*

108 A double pin gauge is a hand tool used for

A *shaping mortises*
B *cutting mortises*
C *cleaning-out mortises*
D *marking-out mortises*

109 A suitable fixing device for securing a hardwood threshold to a concrete step would be

- *A* *wooden plugs*
- *B* *metal plugs*
- *C* *gravity toggles*
- *D* *masonry nails*

110 The vertical members in a non-loadbearing timber partition are referred to as

- *A* *studs*
- *B* *uprights*
- *C* *mullions*
- *D* *jambs*

111 In a felled log the greatest shrinkage will occur

- *A* *at the pith of the trunk*
- *B* *along the length of the trunk*
- *C* *tangentially to the circumference of the trunk*
- *D* *radially to the circumference of the trunk*

112 The top of a timber gate post can be protected from the weather by

- *A* *three coats of paint*
- *B* *weathered hardwood cap*
- *C* *mild steel cap*
- *D* *varnishing*

113 A gravel board is

- *A* *a horizontal board used at eaves level to support the gutter*
- *B* *a horizontal board used at the edge of scaffolding to prevent materials falling off*
- *C* *a horizontal board used in precast formwork*
- *D* *a horizontal board used at the bottom edge of a vertically boarded fence*

114 Which of the following properties is most important for timber used in wood block floors?

A *tensile*
B *abrasion*
C *figure*
D *bending*

115 Partitions are often fitted with glazed lights at their ceiling level. This type of glazing is known as

A *sub-light*
B *fanlight*
C *dead light*
D *borrowed light*

116 Figure 19 shows a vertical section through a light partition at floor level. To carry the weight of the partition the floor would be strengthened by

A *increasing the thickness of the floorboards*
B *fixing solid strutting between the two joists*
C *increasing the depth of the joists*
D *increasing the thickness of the joists*

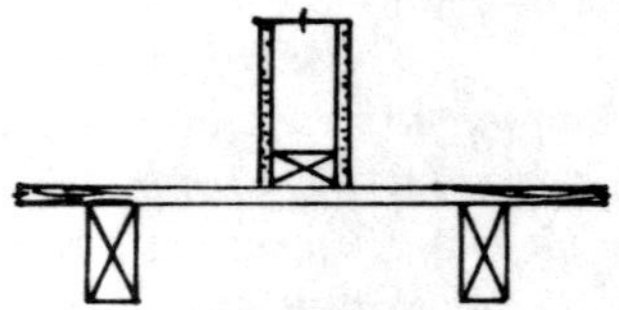

Figure 19

117 When fixing sheets of hardboard to a stud partition, what precautions should be taken to prevent the boards buckling after they are fixed?

A *provide a 3 mm gap between sheets for moisture movement*
B *leave the sheets to acclimatise for three days*
C *damp the back of the sheets with water and leave for two days before fixing*
D *fix the sheets immediately after delivery*

118 Which is a suitable hinge for hanging the gate shown in Figure 20?

A *band and gudgeon*
B *150 mm light tie hinge*
C *75 mm right hand rising butt*
D *100 mm nylon washered brass butt*

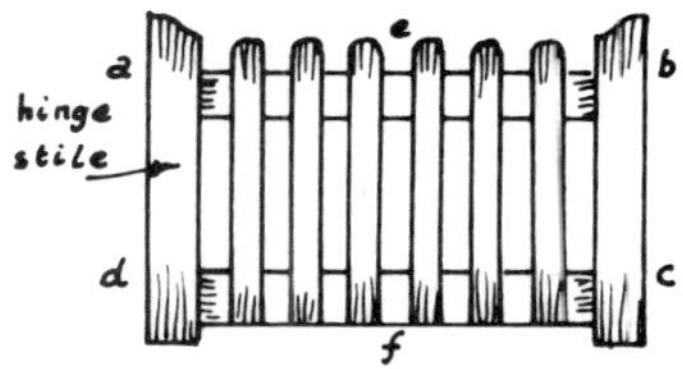

Figure 20

119 The gate in Figure 20 requires a brace which should be fixed between

A *a and c*
B *d and b*
C *f and a*
D *e and c*

120 Figure 20 shows a

A *trellised gate*
B *open board gate*
C *baluster gate*
D *palisade gate*

121 Asbestos is a material which is particularly useful as a

A *good sound insulator*
B *very good fire resistant*
C *damp proof barrier*
D *bath panel*

122 The intersection at an internal angle of the skirting shown in Figure 21 would be jointed by a

A *mitre and scribe*
B *plain mitre*
C *scribe*
D *mason mitre*

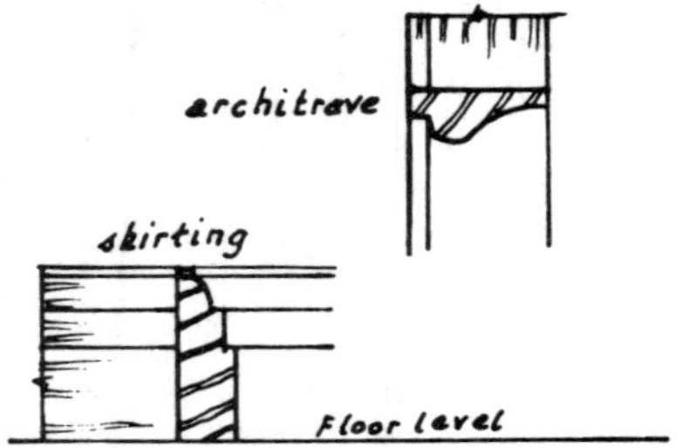

Figure 21

123 The intersection of the skirting and architrave in Figure 21 would be jointed by

A *square cutting to skirting and architrave and fixing both to a plinth block*
B *mitring the skirting to the centre of the architrave*
C *scribing the skirting over the architrave*
D *chamfering the end of the skirting to the architrave*

124 The moulding in Figure 22 is known as

A *a cyma recta*
B *a cyma reversa*
C *a pencil round*
D *an ovolo*

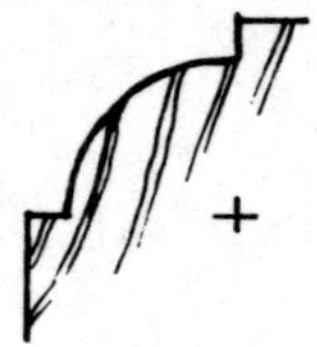

Figure 22

125 2·25 ÷ 0·75 is

A *1·125*
B *0·75*
C *0·33*
D *3·0*

126 Self-tapping screws are used for fixing timber to

A *sheet steel*
B *wood battens*
C *concrete floors*
D *corrugated asbestos*

127 A site-fixing cartridge gun operates by using the principle of

A *electronics*
B *percussion*
C *pneumatics*
D *the computer*

128 A 2-metre-long site-levelling board when checked for accuracy is found to be 5 mm out of parallel. If used in this state over a distance of 8 metres the inaccuracy will be

A *40 mm*
B $\frac{5}{8}$ *mm*
C *80 mm*
D *20 mm*

129 A softwood chamfered architrave ex 75 by 25 mm is jointed at its intersection between jamb and head by a

A *tongue and groove*
B *mitre*
C *dowel*
D *mortise and tenon*

130 In Figure 23 how many angles are equal to 135°?

A *4*
B *6*
C *8*
D *10*

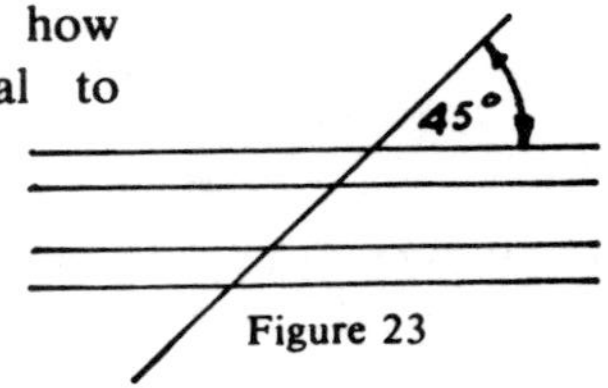

Figure 23

131 The accuracy of a spirit level and a 1 metre straight-edge can be checked by *reversing*

A *the level and straight-edge*
B *the level only*
C *the straight-edge only*
D *the level only at each end*

132 Figure 24 shows a

A *shear plate*
B *corner plate*
C *framing anchor*
D *grid anchor*

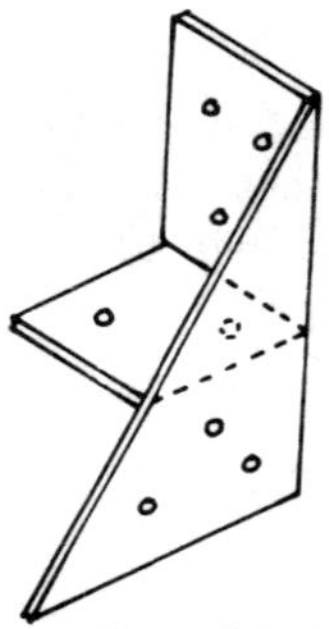

Figure 24

133 The device shown in Figure 24 is used

A *for strengthening the corners of a packing case*
B *for repairing damaged sashes*
C *for lengthening a ridge board*
D *as a fixing device for jointing carpentry timbers*

134 The twist on a wooden pallet plug can be cut by using

A *a handsaw*
B *a carpenter's axe*
C *a 6 mm firmer chisel*
D *a jack plane*

135 The feet of an external door-frame, without a threshold, can be secured to a concrete step by

A *75 mm × 13 mm diameter galvanised iron dowel*
B *two 100 mm oval wire nails*
C *bedding into a recess in the concrete*
D *using a W1 angle cleat*

136 A 100 by 25 mm coat rail can be fixed to a hollow plasterboard dry partition by using

A *a spring toggle bolt*
B *a fibre plug and japanned screw*
C *a self-tapping screw*
D *lost head nails*

137 A site foreman asks you to make him a set of profiles. He will use these

- *A* *for gauging the brickwork courses*
- *B* *as temporary boards supporting a trench*
- *C* *for setting-out the footings of a building*
- *D* *as temporary braces supporting a window frame while it is being built-in*

138 A suitable catch for a ledged and braced batten door is

- *A* *a bales catch*
- *B* *a Norfolk thumb latch*
- *C* *a dead rim lock*
- *D* *A roller mortise latch*

139 Which of the following is classed as second fixing?

- *A* *laying ground floor joists*
- *B* *erecting a timber stud partition*
- *C* *fixing architraves and skirtings*
- *D* *fixing door linings*

140 In Figure 25 the line a–b is termed

- *A* *a chord*
- *B* *a tangent*
- *C* *a normal*
- *D* *a segment*

141 Hardwoods are pre-bored for nails to

- *A* *prevent the wood splitting*
- *B* *ensure the nails are driven in straight*
- *C* *give accurate positioning of the nails*
- *D* *speed up nailing*

142 To determine the centre for the segmental arch in Figure 25, bisect

- *A* *ac and cd*
- *B* *ad and db*
- *C* *cb and dc*
- *D* *ab and ac*

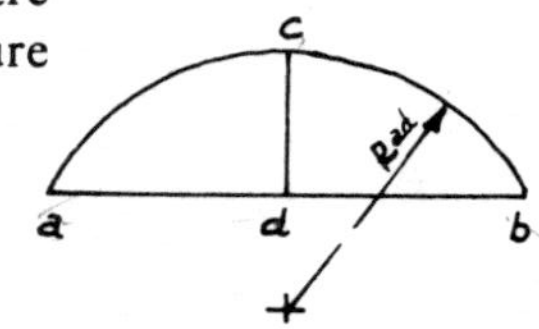

Figure 25

143 The main reason for leaving horns on a door until it is ready for hanging is

- *A* *to permit the scribing of the top and bottom edges*
- *B* *to prevent damage to the door during transit*
- *C* *to permit a short door to be lengthened*
- *D* *to prevent the door twisting during storage*

144 A cambered brick arch of 1 metre span and 100 mm rise is supported during its construction by

- *A* *a semi-circular centre*
- *B* *a framed centre*
- *C* *a turning piece*
- *D* *an easing piece*

145 The easing and striking of an arch centre is done by the use of

- *A* *mould oil*
- *B* *radius rod*
- *C* *string line*
- *D* *folding wedges*

146 The easing and striking of formwork for an in-situ concrete lintel can be done by using

- *A* *metal dogs*
- *B* *adjustable steel props*
- *C* *fox wedges*
- *D* *block wedges*

147 Which of the following applies to formwork moulds for pre-cast concrete units?

- *A* *they enable several identical units to be cast from one mould*
- *B* *they can only be used for special jobs*
- *C* *they are more expensive than in-situ formwork*
- *D* *there is less possibility of timber shrinkage during use*

148 A concrete door canopy has been cast in-situ during cold weather. The formwork should be struck after

A *five days*
B *fourteen days*
C *twenty-one days*
D *twenty-eight days*

149 The top edge of a pre-cast concrete lintel should be clearly marked 'T' to

A *indicate the position of the steel reinforcement*
B *indicate the face side up of the lintel*
C *form a bedding key*
D *form a channel to drain away surplus water*

150 Nails specially made for fixing formwork members have one of the following features

A *sherardised to prevent rusting*
B *double headed for easier withdrawal*
C *made from high tensile steel for greater strength*
D *small diameter to prevent splitting of the timber*

151 Which number gives the highest value when all the figures of the line are added together?

A *4 6 8 2 4 4*
B *5 6 7 8 9 5*
C *3 4 5 7 8 6*
D *4 3 6 8 8 4*

152 How many millimetres are equal to $\frac{3}{8}$ metre?

A *375*
B *325*
C *500*
D *275*

153 Figure 26 shows part of a typical builders scale, known as

A *a closed scale*
B *a diagonal scale*
C *an open divided scale*
D *a surveying scale*

154 In Figure 26 the smallest division x represents

A *1 mm*
B *5 mm*
C *10 mm*
D *15 mm*

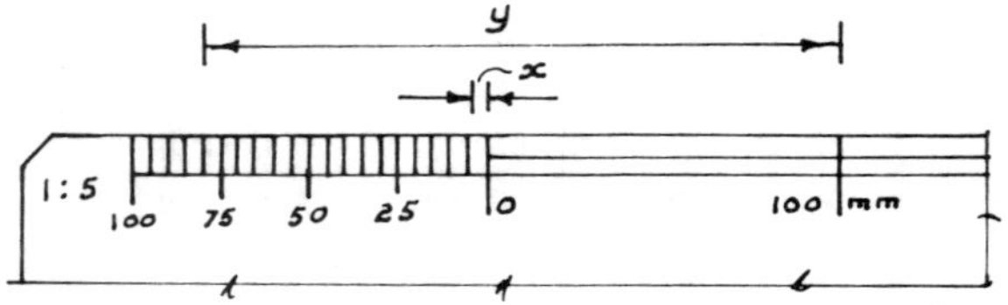

Figure 26

155 The dimension y in Figure 26 reads

A *175·1 mm*
B *1·800 m*
C *80 mm*
D *180 mm*

156 A scale of 50 mm to 1 m has a representive fraction of

A *1 : 50*
B *1 : 20*
C *1 : 5*
D *1 : 200*

157 Dimensions should be written on to a drawing rather than left to be measured by a scale rule, because

A *humidity variations in the atmosphere can change the true value of the dimension*
B *the scale rule could be inaccurate*
C *scale rules are difficult to read*
D *it is difficult to obtain a scale rule*

158 A suitable scale for drawing the elevation of a window 1·200 m wide by 1·350 m high would be

A *1 : 2*
B *1 : 200*
C *1 : 20*
D *1 : 50*

159 A standard reference, which gives detailed information about building drawing office practice, is

A *Building Regulations, part M*
B *TRADA Guide to Building*
C *Standard Method of Measurement*
D *BS 1192*

160 A popular grade of pencil lead for drawing feint construction lines is

A *HB*
B *3B*
C *3H*
D *6H*

161 The scale drawing of a folding wedge shown in Figure 27 is drawn in

A *first angle orthographic*
B *third angle orthographic*
C *isometric*
D *planometric*

162 In Figure 27 the angles P are always

A *15°*
B *30°*
C *45°*
D *20°*

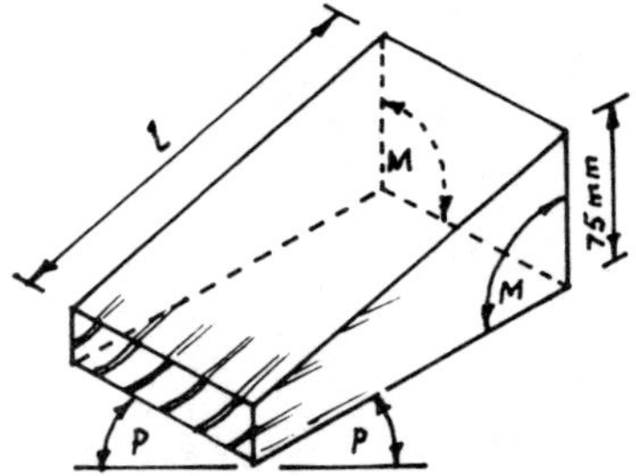

Figure 27

163 In Figure 27 the angles M are always

A *120°*
B *60°*
C *100°*
D *135°*

164 If Figure 27 is drawn to a scale of 1 : 5, the given dimension will actually be drawn as

A *1·5 m*
B *15 mm*
C *25 mm*
D *12·5 mm*

165 In Figure 27 the dimension L cannot be measured from the drawing because

- *A* *it is not on the isometric axis*
- *B* *its dimension is not required*
- *C* *all dimensions in isometric projection are distorted*
- *D* *it doesn't fit a drawing set-square*

166 Figure 28 shows a short length of timber moulding drawn in

- *A* *dimetric*
- *B* *orthographic*
- *C* *oblique*
- *D* *isometric*

167 The angle P in Figure 28 is usually

- *A* *45°*
- *B* *60°*
- *C* *30°*
- *D* *15°*

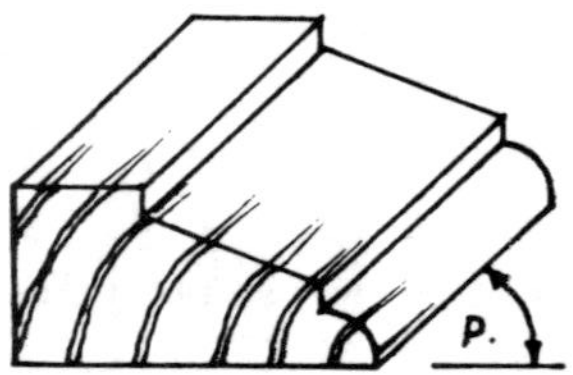

Figure 28

168 When Figure 28 is drawn in cabinet oblique projection, the dimensions of the drawing

- *A* *can be measured in section but not in length*
- *B* *can all be measured*
- *C* *can be measured in length but not in section*
- *D* *are all distorted*

169 The moulding in Figure 28 costs 10p per metre lineal, therefore 260 metre run will cost

- *A* *£2·60*
- *B* *£260·00*
- *C* *£26·00*
- *D* *260p*

170 Which is the accepted method of drawing a dimension line and writing the dimension in metric units.

A 

B 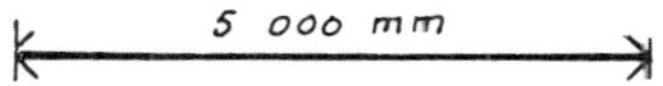

C 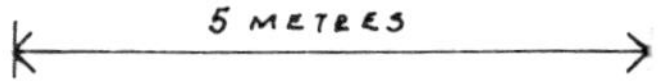

D 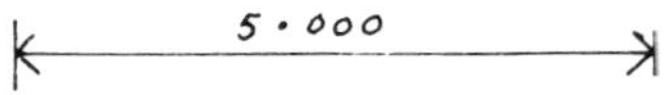

171 The timber threshold shown in Figure 29 is drawn in

A *orthographic, first angle projection*
B *orthographic, third angle projection*
C *photographic projection*
D *axonometric projection*

172 The line x–y in Figure 29 indicates

A *a base line*
B *intersection of VP with HP*
C *a data line*
D *a view plane*

173 The vertical section P–P in Figure 29 is shown by

A *view O*
B *view N*
C *view M*
D *view L*

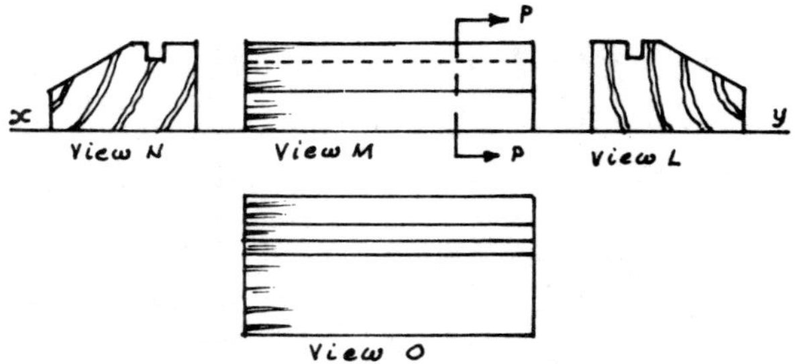

Figure 29

174 On a drawing the centre line of an object is indicated by

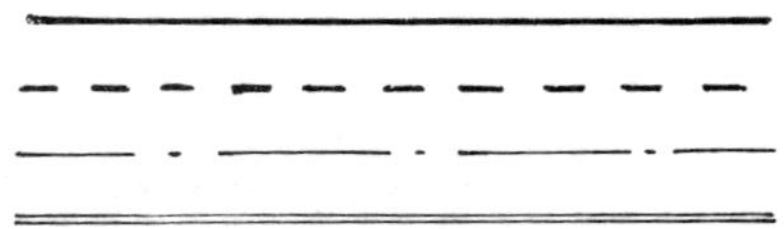

A
B
C
D

175 In passing on a customer's requirements for a range of special kitchen cupboards, the best means of communication would be

A *written statements*
B *freehand sketches*
C *verbal instruction*
D *scaled drawings*

176 Thrust wheels are always fitted to one of the following machines

A *wood-turning lathe*
B *narrow bandsaw*
C *chisel mortiser*
D *pendulum crosscut saw*

177 On completion of turning a timber baluster, it is separated from the chucks by the use of a

A *diamond chisel*
B *bullnose chisel*
C *parting chisel*
D *vee chisel*

178 The blade of a narrow bandsaw is kept running true by

A *frequent sharpening and setting of the saw teeth*
B *adjustment of the guide rollers*
C *cutting straight-grained timber only*
D *use of tracking and tensioning device*

179 A 450 mm diameter sanding-disc is to be used on a wood turning lathe. It is secured to the lathe spindle by

A *an outside face plate*
B *concentric chuck*
C *drive chuck*
D *coric chuck*

180 The nett length of a saw blade for the narrow bandsaw in Figure 30 is calculated from the formula

A $\pi r^2 + 2W$
B $2\pi r + W$
C $\pi r + 2W$
D $2\pi r + 2W$

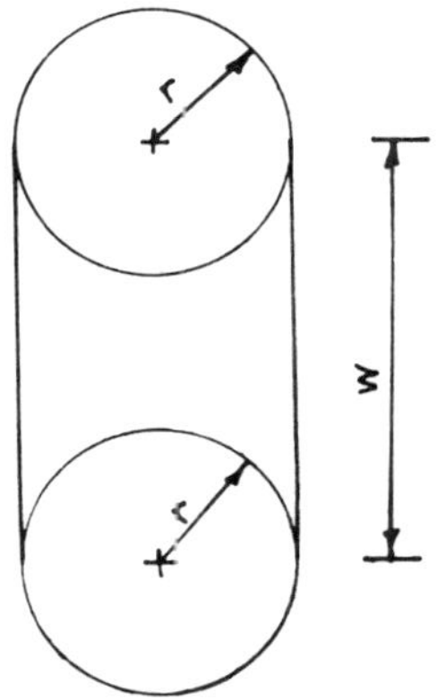

Figure 30

181 A 3-pin 13-amp plug must have the new colour code wires connected as follows:

	Live pin	Neutral pin	Earth pin
A	*brown*	*yellow/green*	*blue*
B	*blue*	*yellow/green*	*brown*
C	*brown*	*blue*	*yellow/green*
D	*blue*	*brown*	*yellow/green*

182 The cutting action of an overhand planer works upon one of the following principles

A *rotary*
B *shearing*
C *reciprocating*
D *slicing*

183 A metal fore or wood try plane is used for

A *smoothing surfaces to a fine finish*
B *trueing large surfaces to a flat plane*
C *jacking off sawn surfaces*
D *general purpose planing*

184 The floor finish around a woodworking machine should preferably be

- *A* *tongued and grooved softwood floorboards*
- *B* *sharp sand and cement*
- *C* *hardwood blocks set in bitumen*
- *D* *non-slip composite flooring*

185 An acceptable temperature for comfortable and safe working in a joiner's shop is in the region of

- *A* *10°C–16°C (50°F–60°F)*
- *B* *8°C–14°C (45°F–55°F)*
- *C* *6°C–12°C (40°F–50°F)*
- *D* *16°C–20°C (60°F–68°F)*

186 A riving knife must be always used with a circular saw machine in order to

- *A* *clear the sawdust from the timber*
- *B* *prevent binding of the timber on the back of the saw*
- *C* *scrape a good finish to the sawn face of the timber*
- *D* *steady the timber as it clears the saw blade*

187 The crown guard on a circular saw machine is provided to

- *A* *cover the unused part of the saw blade above the saw bench*
- *B* *reduce noise pollution*
- *C* *cover the unused part of the saw blade below the saw bench*
- *D* *prevent kick-back of the sawn timber*

188 A suitable circular saw for general purposes use on a joiners sawbench would be

- *A* *swage saw*
- *B* *taper saw*
- *C* *ground-off saw*
- *D* *parallel plate saw*

189 When hand-feeding a circular saw bench the push stick should be used to

A *tighten the saw blade packings*
B *keep the operator's hands clear of the saw blade*
C *prevent the timber from tipping up*
D *guide the timber through the saw*

190 When ripping timber on a circular sawbench the front end of the fence should be parallel to

A *the gullet of the saw teeth at the front*
B *the riving knife at the rear*
C *the centre of the saw blade*
D *mid-way between the centre and front of the saw*

191 The chief feature which affects the density of a seasoned hardwood is

A *size of pores*
B *amount of wood substance*
C *age of the tree*
D *amount of sapwood*

192 Which of the following seasoned timbers would be expected to have a high rate of moisture absorption?

A *Greenheart*
B *English oak*
C *Lignum vitae*
D *Poplar*

193 The three physical states of matter are

A *solid, vapour and gas*
B *vapour, liquid and gas*
C *solid, liquid and gas*
D *solid, liquid and vapour*

194 Which of the following materials would be classed as porous?

A *vitrolite tile*
B *common brick*
C *lead pipe*
D *plastic door handle*

195 Which is correct

A $density = \frac{mass}{volume} \times 100$

B $density = \frac{specific\ gravity}{volume} \times 10$

C $density = \frac{specific\ gravity}{mass}$

D $density = \frac{mass}{volume}$

196 The clearance gap between a riving knife and the teeth of a circular saw must not exceed

A *15 mm*
B *12 mm*
C *20 mm*
D *30 mm*

197 The cutterblock guard on a surface planer is known as

A *crown guard*
B *cage guard*
C *bridge guard*
D *Shaw guard*

198 Which of the following operations requires *special safeguard* when being worked on the overhand planer

A *rebating*
B *chamfering*
C *levelling*
D *flat surfacing*

199 Unevenly spaced ripple marks on machine-planed timber can be caused by

A *cutters not operating in the same cutting circle*
B *blunt cutting knives*
C *feed speed too slow*
D *hard and soft grain of timber*

200 Holding-down springs are used as a

A *balancing mechanism on the arm of a mortiser*
B *tensioning device on a bandsaw*
C *pressure device on a lathe*
D *safety device on a surface planer*

201 An acceptable number of cuttermarks per 25 mm for internal joinery is

A *10*
B *15*
C *20*
D *30*

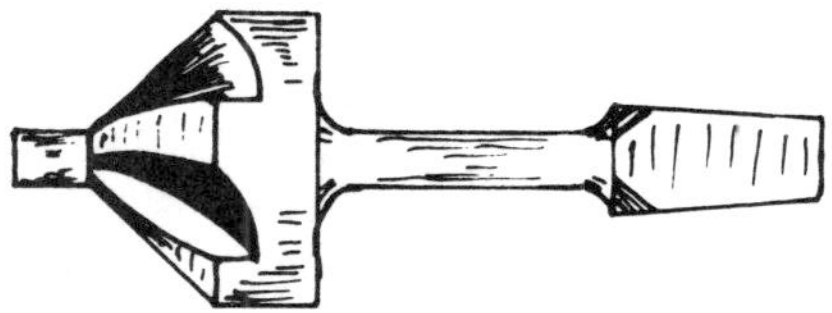

Figure 31

202 The bit shown in Figure 31 is used in a joiner's sweep brace for

A *countersinking for screws*
B *boring blind holes*
C *reamering existing holes*
D *sharpening square chisels*

203 The standard range of chisel and auger sizes for a mortising machine which are available from stock are

A *6 mm–25 mm*
B *10 mm–25 mm*
C *6 mm–38 mm*
D *10 mm–38 mm*

204 The cutter head of a single mortising machine is brought into contact with the timber by

A *foot operated rise and fall table*
B *use of feed rollers*
C *a lever handle*
D *feeding the timber by hand pressure*

205 The gap between the ends of the auger and chisel on a mortising machine should be

A *1·5 mm*
B *3 mm*
C *4 mm*
D *nil*

206 The nut on a circular saw spindle is prevented from working loose during use by the

A *location pin through the saw blade*
B *vee thread on the nut and spindle*
C *spring washer*
D *left-hand thread on the mandrell*

207 The SI unit multiple which is associated with the prefix 'mega' has a value of

A *1 000 000*
B *1000*
C *100*
D *10*

208 A load of timber contains 4 cubic metres and weighs 2800 kg. The density of the timber is

A $70\ kg/m^3$
B $140\ kg/m^2$
C $700\ kg/m^3$
D $350\ kg/m^3$

209 Thresholds to external door frames should be made from hardwood, free of sapwood and with

A *a high density*
B *a medium density*
C *a low density*
D *any density*

210 Balsa wood has a very low density because it

A *is a softwood*
B *contains a large number of voids*
C *is a very dry timber*
D *contains a high percentage of natural oils*

211 Which adhesive is a by-product of animal waste from abattoirs?

A *synthetic resin*
B *scotch glue*
C *casein cement*
D *impact*

212 Three-core plywood is normally available up to a maximum thickness of

A *9 mm*
B *12 mm*
C *15 mm*
D *18 mm*

213 The setting action of casein adhesive depends upon

A *evaporation of water*
B *minimum pressure of 7000 kg/m²*
C *a chemical reaction*
D *a temperature of 21°C*

214 A timber merchant can supply BS design and sizes of architrave, skirtings, etc., from stock. These are known as

A *trim*
B *cover fillet*
C *stock mouldings*
D *bed mouldings*

215 Which adhesive cannot be used for external work?

A *casein cement*
B *synthetic resin*
C *epoxy resin*
D *scotch glue*

216 A timber is classified as a hardwood because

A *of its density*
B *of the arrangement and structure of its cells*
C *it is available in widths over 450 mm*
D *its tree grows only in tropical regions*

217 Which adhesive is suitable for bonding laminated plastic sheets to a laminboard worktop?

A *latex contact adhesive*
B *soya bean paste*
C *poly-vinyl acetate glue*
D *casein cement*

218 The most popular softwood used in Britain for joinery work is

A *Western hemlock*
B *Yellow pine*
C *Pitch pine*
D *European redwood*

219 Which adhesive will produce a dark stain on certain timbers?

A *P.V.A*
B *casein*
C *gelatine*
D *phenol formaldehyde*

220 A timber merchant will normally base his price for sheet materials such as plywood, hardboard, etc., on a unit quantity price per

A $1000\ m^2$
B $100\ m^2$
C $10\ m^2$
D $1\ m^2$

221 A synthetic resin adhesive requires the addition of a catalyst to

A *make it easier to work*
B *prevent staining of the wood*
C *make the glue line transparent*
D *produce a chemical reaction*

222 Scots pine is imported into Britain from

A *Scandinavia*
B *Canada*
C *Western Africa*
D *South America*

223 A pot of casein adhesive has stood unused for several hours. Before further use it should be

- *A* *vigorously re-whisked*
- *B* *thrown away and a fresh mix prepared*
- *C* *re-mixed with cold water only added*
- *D* *re-mixed with powder only added*

224 Which timber will cause corrosion of iron fittings when fixed in a damp situation?

- *A* *Mahogany*
- *B* *Oak*
- *C* *Beech*
- *D* *Sycamore*

225 Exterior-grade plywood is bonded with

- *A* *soya glue*
- *B* *casein cement*
- *C* *blood albumen*
- *D* *synthetic resins*

226 Prime oak boards show a conspicuous 'silver figure' on their face sides. This is obtained by

- *A* *quarter sawing*
- *B* *through and through sawing*
- *C* *flat sawing*
- *D* *tangential sawing*

227 Which timber is classified as a hardwood?

- *A* *South American parana*
- *B* *European redwood*
- *C* *European red beech*
- *D* *Western red cedar*

228 Plywood can be marked 'WBP' to indicate it is

- *A* *weather and boil proof*
- *B* *water and boil proof*
- *C* *weather and beetle proof*
- *D* *water and beetle proof*

229 Which of the following abrasive papers is most suitable for finishing off a softwood door frame?

- *A* *emery paper*
- *B* *flour paper*
- *C* *flint paper*
- *D* *glass paper*

230 A centre bit for use in a sweep ratchet brace is particularly useful for boring

- *A* *deep holes up to 20 mm diameter*
- *B* *shallow holes over 20 mm diameter*
- *C* *blind holes any diameter*
- *D* *holes for 6 mm diameter wood dowels*

231 The glued width joint shown in Figure 32 is made by using

- *A* *spindle moulder machine*
- *B* *wood moulding plane*
- *C* *metal combination plane*
- *D* *metal chamfer plane*

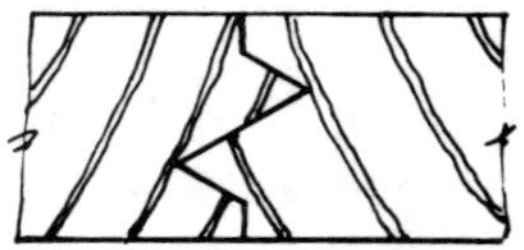

Figure 32

232 19 mm tongued and grooved boards used in framed, ledged and braced doors usually have their meeting edges

- *A* *finished by a planted bead*
- *B* *finished by a stuck ovalo*
- *C* *finished by a stuck chamfer*
- *D* *left as square edged*

233 Figure 33 shows the section for a 63 × 88 mm jamb and head to an external door frame. A suitable size for the thickness of a single tenon would be

- *A* *10 mm*
- *B* *19 mm*
- *C* *28 mm*
- *D* *32 mm*

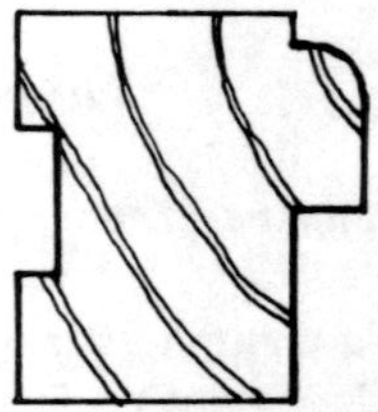

Figure 33

234 Two hundred door frames, with section as Figure 33, are to be made in one batch. At the joint between jamb and head the moulding would be

- *A* *mitred*
- *B* *bench-scribed*
- *C* *through-scribed*
- *D* *mason-mitred*

235 The groove at the back of the frame in Figure 33 is provided

- *A* *for bedding the frame to the wall*
- *B* *to prevent warping of the timber*
- *C* *as a channel for draining away rainwater*
- *D* *to receive metal anchor straps*

236 Storey frames are normally used for door openings in

- *A* *external cavity walls*
- *B* *timber stud partitions*
- *C* *load-bearing brick partitions*
- *D* *lightweight block partitions*

237 Which is the best shape in Figure 34 for a wedge to be used in the mortise and tenon joint of a casement sash?

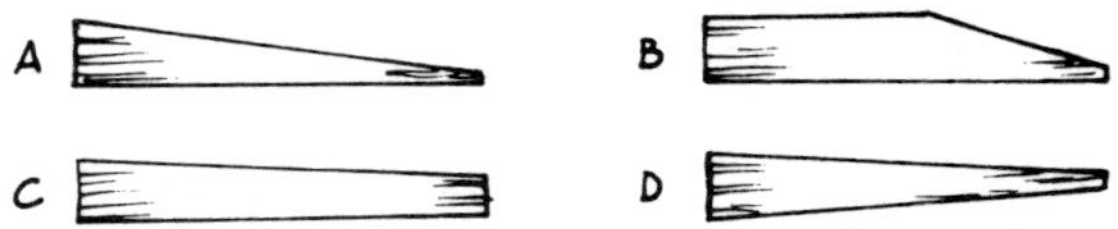

Figure 34

238 Door and window frames are primed before site delivery to

- *A* *hide the knots and fill up cracks*
- *B* *prevent absorption of moisture*
- *C* *provide an undercoat for the painter*
- *D* *give the timber a distinctive colour*

239 A timber merchant's abbreviation 'P.S.E.' means

- *A* *planed square edge*
- *B* *prime square edge*
- *C* *planed softwood ends*
- *D* *prime sawn ends*

240 The horn or lug of a door-frame head is cut as shown in Figure 35 when it is built into a

- *A* *timber stud partition*
- *B* *fair face brick wall*
- *C* *breeze slab partition*
- *D* *concrete block wall*

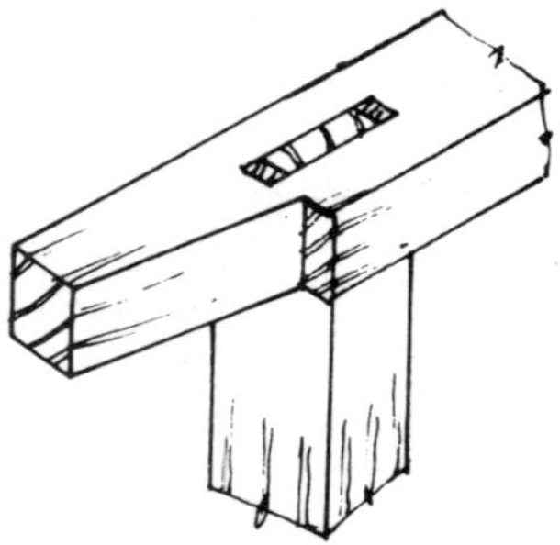

Figure 35

241 Figure 36 shows the rear elevation of a door commonly used for outbuildings. It is known as a

- *A* *framed, ledged and braced door*
- *B* *matchboarded door*
- *C* *ledged and braced batten door*
- *D* *ledged batten door*

242 The distance x in Figure 36 should not exceed

- *A* *50 mm*
- *B* *75 mm*
- *C* *100 mm*
- *D* *150 mm*

243 The distance y in Figure 36 should not be less than

- *A* *16 mm*
- *B* *19 mm*
- *C* *25 mm*
- *D* *38 mm*

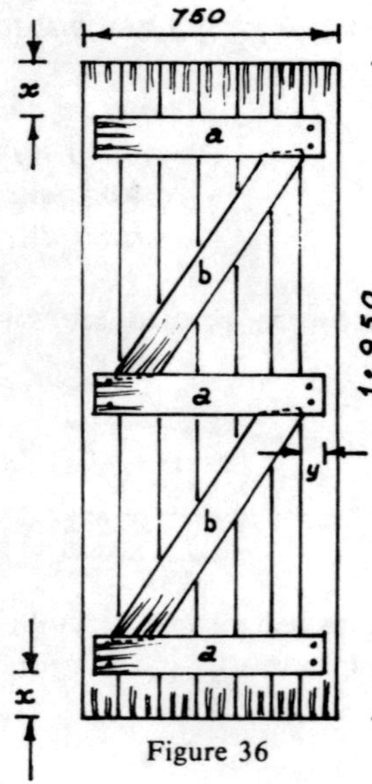

Figure 36

244 The minimum nominal section size for the members a and b in Figure 36 is

A *25 × 100 mm*
B *25 × 75 mm*
C *31 × 100 mm*
D *25 × 150 mm*

245 The door in Figure 36 requires seven boards. The total length of boards, including 10% for waste, will be

A *13·650 m*
B *15·015 m*
C *14·650 m*
D *16·015 m*

246 A casement window can be described as a

A *fixed sash in a timber sub-frame*
B *solid frame fitted with hinged opening sashes*
C *cased frame fitted with balanced sliding sashes*
D *solid frame fitted with balanced sliding sashes*

247 Which of the mouldings in Figure 37 is not suitable for scribing?

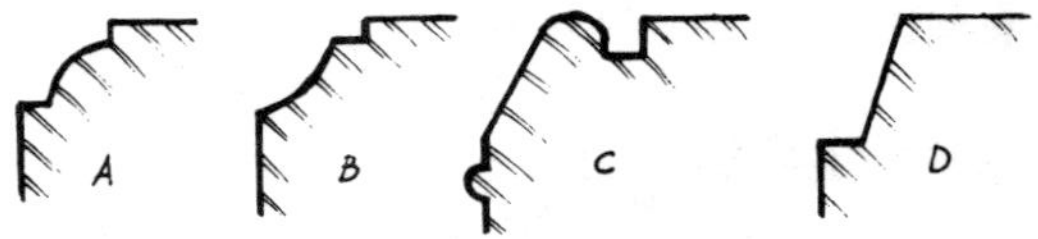

Figure 37

248 The horizontal member placed between the head and cill of a casement window is known as a

A *muntin*
B *mullion*
C *transome*
D *drip-rail*

249 A single light casement window with one side-hung outward-opening sash will require the following ironmongery

- *A* *cockspur fastener and two-pin stay*
- *B* *one-pin stay*
- *C* *roller catch and two-pin stay*
- *D* *cockspur fastener and quadrant stay*

250 Planted rebate stops to internal door casings are fixed

- *A* *before hanging the door*
- *B* *after hanging the door*
- *C* *after hanging the door and fixing the latch*
- *D* *after the first undercoat*

251 M.O.S. is an abbreviation for

- *A* *mitre or scribe*
- *B* *moulded on the solid*
- *C* *middle or side*
- *D* *mitre over stopped end*

252 A bare-faced tenon could be used when making

- *A* *framed, ledged and braced door*
- *B* *ledged and braced batten door*
- *C* *four panel internal door*
- *D* *flush door*

253 In Figure 38 the moulding M is known as a

- *A* *bolection moulding*
- *B* *solid moulding*
- *C* *panel moulding*
- *D* *bed moulding*

Figure 38

254 The moulding M in Figure 38 is fixed by

A *panel pins to the panel*
B *panel pins to the stile*
C *glue and panel pins*
D *slot screws to the panel*

255 In a panelled door the 10 mm plywood panels are retained in their grooves by

A *glue*
B *panel pins*
C *left-loose hand fit*
D *malleting a tight fit*

256 Which of the timber sections in Figure 39 would give the best arrangement of grain for a 32 × 75 mm cupboard door framing

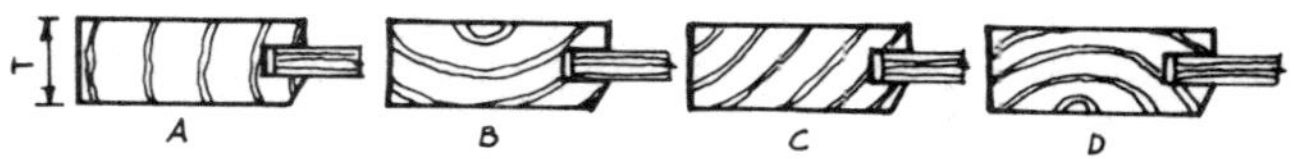

Figure 39

257 A suitable thickness for the tenons in framing the door sections in Figure 39, expressed as a proportion of the thickness T, would be

A $\frac{1}{4}$
B $\frac{1}{3}$
C $\frac{3}{8}$
D $\frac{1}{2}$

258 In a mortise and tenon joint, for window frame, wedge room should be provided for

A *at the sides of the tenon*
B *at the ends of the mortise*
C *at the sides of the mortise and tenon*
D *by saw cuts along the length of the tenon*

259 A suitable depth for a kitchen wall cabinet to BS 1195 is

A *200 mm*
B *400 mm*
C *300 mm*
D *500 mm*

260 The sides of a groove require easing to receive a plywood panel. The plane for this work would be

A *a side rebate*
B *a shoulder*
C *a bullnose*
D *a badger*

261 The area of a circle is calculated by using the formula

A
B πr^2
C $2\pi r$
D πd

262 The area of a circle with a radius of 7·000 metres is

A *154* m^2
B *22* m^2
C *77* m^2
D *112* m^2

263 The part of a circle shown in Figure 40 is known as a

A *segment*
B *normal*
C *tangent*
D *section*

264 The dotted line M–N in Figure 40 is termed

A *a chord*
B *an arc*
C *an inner normal*
D *a radius*

265 If Figure 40 is part of a complete circle having an area of 240 m^2, the area of Figure 40 will be

A *120 m^2*
B *40 m^2*
C *80 m^2*
D *60 m^2*

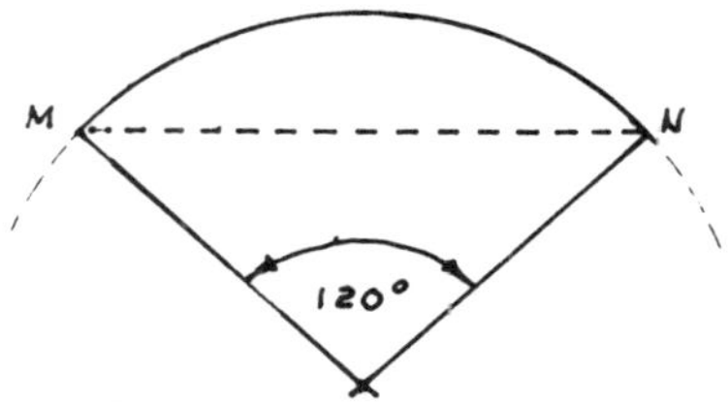

Figure 40

266 A slot-screw joint could be used for

A *making a width joint in shelving*
B *fixing a plywood face to skeleton framing*
C *fixing a sink unit to the floor*
D *forming a track for sliding doors*

267 62 mm pressed steel hinges would not be suitable for hanging bathroom cabinet doors because they are

A *not strong enough to carry the weight of the doors*
B *too large in size for this purpose*
C *liable to corrosion in the high humidity*
D *not face-fixing*

268 Slatted shelving is used in certain types of cupboard because it

A *is cheaper than solid shelves*
B *allows for free circulation of air*
C *is more easily removable*
D *uses up short lengths of timber*

269 The correct order for driving the wedges in the mortise and tenon joint in Figure 41 is

A *2, 4, 1 and 3*
B *4, 1, 3 and 2*
C *1, 2, 3 and 4*
D *3, 2, 4 and 1*

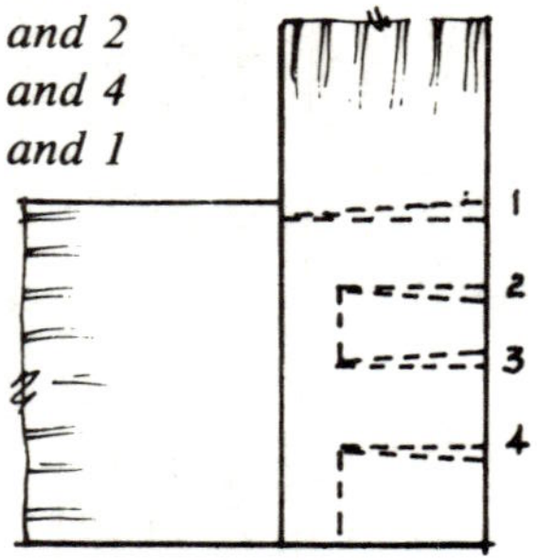

Figure 41

270 In slatted shelving the slats should be fixed to their bearers by

A *corrugated fasteners*
B *oval wire nails*
C *countersunk brass screws*
D *star dowels*

271 In the mortise and tenon joint shown in Figure 42 the glue would be applied to the areas

A *N*
B *C*
C *O*
D *N, C and O*

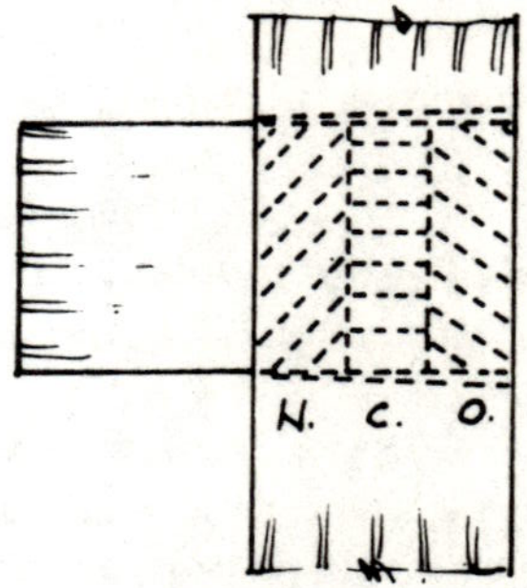

Figure 42

272 A method of setting-out a true elliptical curve on the site would be to use a

A *trammel*
B *radius rod*
C *protractor*
D *steel square*

273 Figure 43 shows four equal-size circles within a square. This is known as a pattern of

A *circumscribed circles*

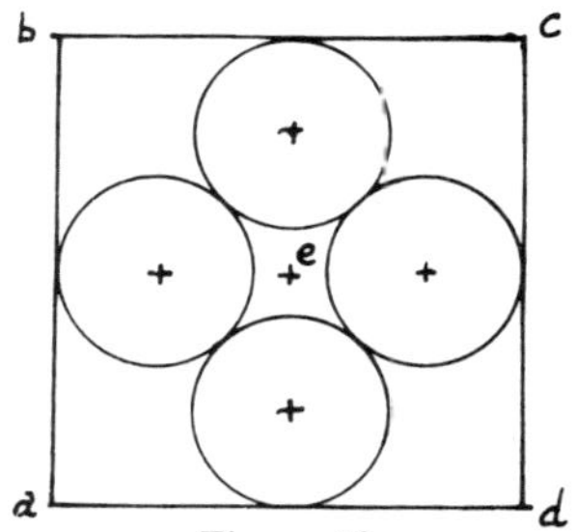

Figure 43

B *eccentric circles*
C *inscribed circles*
D *trefoil circles*

274 The centres for the circles in Figure 43 are obtained by bisecting

A *angles d–a–b and b–c–a*
B *angles e–a–b and a–b–e*
C *angles b–c–d and b–c–e*
D *angles a–d–c and c–b–d*

275 The diameter of a circle with a circumference of 220 mm would be

A *35 mm*
B *70 mm*
C *140 mm*
D *80 mm*

276 Figure 44 shows the plan and elevation of a right circular prism with the top surface inclined at 45° to the horizontal. Which of the following shows the development of the vertical surfaces?

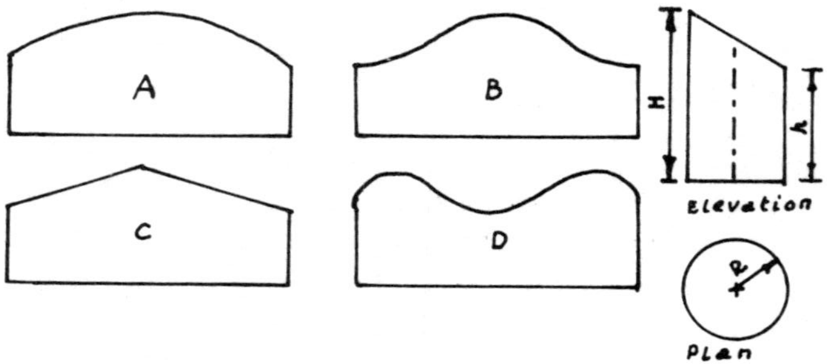

Figure 44

277 The true shape of the top surface in Figure 44 would be

A *circular*
B *diamond-shaped*
C *elliptical*
D *egg-shaped*

278 The formula for calculating the vertical surface area in Figure 44 is

A $\pi r H h$
B $2\pi r (H + h)$
C $\pi r (H + h)$
D $\pi r \frac{1}{2} (H + h)$

279 A lapped dovetail joint could be used in making

A *panelled doors*
B *casement window*
C *kitchen drawers*
D *door casing*

280 BS specification recommends the height of a kitchen worktop level shall be

- *A* *1·000 m*
- *B* *900 mm*
- *C* *850 mm*
- *D* *800 mm*

281 What formula is used to calculate the circumference of an ellipse, when a and b are the semi-minor and semi-major axes?

- *A* πab
- *B* $\pi(a + b)$
- *C* $\pi\frac{1}{2}\,(a + b)$
- *D* $\pi\frac{1}{2}\,(b \div a)$

282 When using contact adhesive the two surfaces should be joined together

- *A* *immediately after the adhesive has been spread*
- *B* *when the adhesive changes from light to dark colour*
- *C* *when the adhesive is touch dry*
- *D* *when the strong odour of the adhesive has disappeared*

283 Pocket screwing is a method of fixing

- *A* *a table top to its supporting rails*
- *B* *a sash pocket to the pulley stile*
- *C* *a hardwood architrave to a softwood ground*
- *D* *a removable display frame*

284 A suitable moisture content for timber to be used in making kitchen furniture is

- *A* *8–10%*
- *B* *11–13%*
- *C* *14–16%*
- *D* *17–20%*

285 A suitable pitch or slope for drawer dovetails is

A *1 in 10*
B *1 in 8*
C *1 in 6*
D *1 in 4*

286 Figure 45 shows a trammel for setting-out the circumference of an ellipse. The distances a and b are equal to

A *the minor and major axes*
B *two thirds the minor and major axes*
C *half the minor and major axes*
D *one third the minor and major axes*

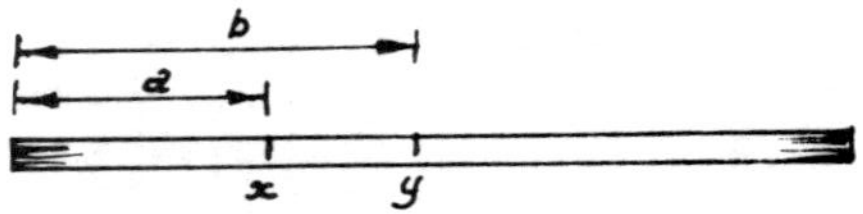

Figure 45

287 When using the trammel in Figure 45 the points x and y are positioned on the

A *major axis only*
B *major and minor axes*
C *minor axis only*
D *foci points*

288 Which part of Figure 46 illustrates concentric circles?

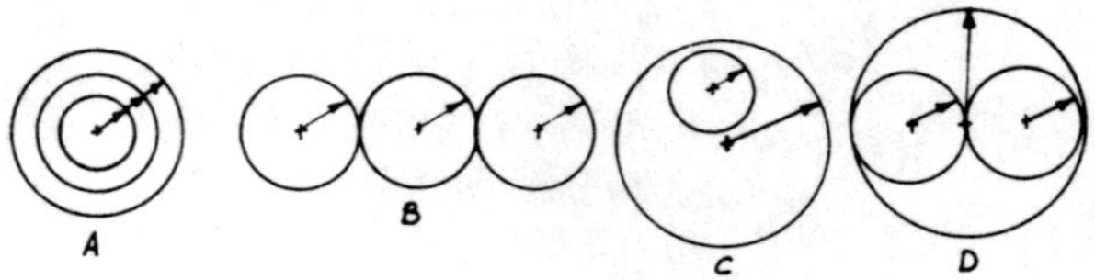

Figure 46

289 A false ellipse is set out by using

A *a trammel as shown in Figure 45*
B *a radius rod from different centres*
C *string stretched around the foci points*
D *intersecting ordinates from auxiliary circles*

290 Which hand tool would be used to clean out and ensure a uniform depth to a dado trench?

A *duplex rabbet plane*
B *shoulder plane*
C *router plane*
D *skew rebate plane*

291 Which timber would be most suitable for a good quality draining board?

A *Teak*
B *Oak*
C *Elm*
D *Parana pine*

292 A joiner's 300 mm diameter sawbench can be fixed to the floor by adhesive instead of bolts. The advantage of this method of fixing is

A *it is stronger than bolts*
B *it reduces vibration and noise*
C *it provides an electrical earth*
D *it beds the machine on to an irregular floor surface*

293 Figure 47 shows the end of a softwood plank. The growth rings indicate it has been

A *rift sawn*
B *radially sawn*
C *flat sawn*
D *quarter sawn*

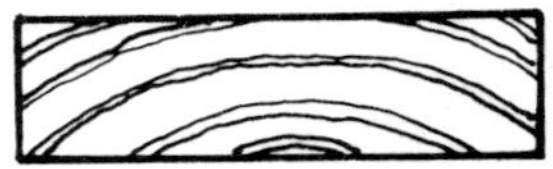

Figure 47

294 On which machine must the riving knife be checked for correct position before using the machine?

- *A* *circular saw*
- *B* *bandsaw*
- *C* *jigsaw*
- *D* *mortiser*

295 Figure 48 shows a triangle circumscribed by a circle. The centre of the circle is obtained by bisecting

- *A* *two angles*
- *B* *two sides*
- *C* *one angle and one side*
- *D* *two angles and one side*

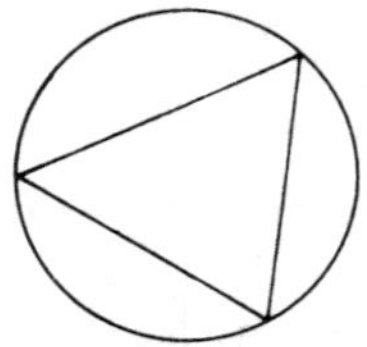

Figure 48

296 A suitable peripheral speed for a circular rip saw machine is

- *A* *20 m/s*
- *B* *30 m/s*
- *C* *40 m/s*
- *D* *50 m/s*

297 A portable site hut, in order to be light but strong, should be constructed

- *A* *on the site with hardwood framing*
- *B* *in sectional units with softwood framing*
- *C* *in softwood framing covered with chipboard sheet externally*
- *D* *as a log cabin with 75 × 200 mm planks*

298 Which is the most suitable timber for larder shelves?

- *A* *Western red cedar*
- *B* *Parana pine*
- *C* *Hickory*
- *D* *Rosewood*

299 A square cutter block must not be used in an overhand planing machine because

A *it revolves too fast*
B *the cutters are difficult to re-sharpen*
C *the cutter ripple marks are widely spaced*
D *it is not permitted by the Woodworking Machine Regulations*

300 A gap-filling glue will provide a strong joint when the glue line is not thicker than

A *1 mm*
B *2 mm*
C *3 mm*
D *4 mm*

301 A working platform on a scaffold, which also has to accommodate materials, should have a minimum width of

A *635 mm*
B *900 mm*
C *1·060 m*
D *1·300 m*

302 An important safety device fitted to portable electric circular saws is

A *a retractable guard*
B *a single-phase electric motor*
C *an adjustable fence*
D *a trigger-type on/off switch*

303 Timber which is to be in constant contact with water should have its preservative applied by

A *hot and cold steeping method*
B *empty cell pressure treatment*
C *full cell pressure treatment*
D *cold dipping method*

304 You are asked to make ten 'darbys' for a fellow craftsman. They will be used by a

A *bricklayer for carrying mortar*
B *plumber for plugging disconnected water pipes*
C *wood machinists in setting-up a double-end tenoner*
D *plasterer as a floating rule*

305 The area of the parallelogram in Figure 49 is

A $27 \cdot 50\ m^2$
B $37 \cdot 21\ m^2$
C $55 \cdot 81\ m^2$
D $18 \cdot 60\ m^2$

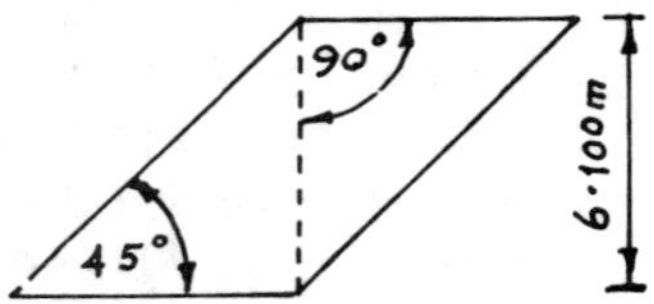

Figure 49

306 An independent scaffold is one which

A *has a double row of standards*
B *is independent of the building for support*
C *is mounted on castor wheels for free movement*
D *can be used in the centre of a floor area*

307 An electric drill with a 12 mm diameter chuck will bore holes in timber to a maximum diameter of

A *12 mm*
B *15 mm*
C *20 mm*
D *25 mm*

308 Which of the following would be classed as a natural defect when selecting timber?

A *pitch pocket*
B *sloping grain*
C *cupping*
D *sap stain*

309 The spores of dry rot fungus (*Merulius lacrymans*) require four basic conditions in order to germinate. These conditions are

A *darkness, water, fresh air and warmth*
B *food, water, darkness and warmth*
C *darkness, food, fresh air and warmth*
D *food, water, fresh air and warmth*

310 A definition of rift-sawn timber is where the growth rings meet the face of the timber at an angle of

A *more than 45°*
B *less than 45°*
C *less than 30°*
D *less than 15°*

311 A fungus which can produce its own moisture is

A *sap-stain*
B *dry rot*
C *wet rot*
D *white rot*

312 A cracked ladder stile should be repaired by

A *replacing with a complete new stile*
B *filling the crack with exterior grade glue and applying cramp pressure*
C *reinforcing with timber straps nailed to the stile*
D *reinforcing with wrought iron straps screwed to the stile*

313 The torque pressure on an electric portable screw-driver is controlled by

A *the voltage supply*
B *hand pressure on the tool*
C *a slip clutch*
D *a two-speed motor*

314 Wire nails can be bought in sacks of 28 lb. The nearest SI equivalent is

- *A* *9·25 kg*
- *B* *10·00 kg*
- *C* *11·20 kg*
- *D* *12·70 kg*

315 The larvae of the Lyctus or powder post beetle will normally be found in

- *A* *sapwood of hardwoods*
- *B* *heartwood of softwoods*
- *C* *multiplywood*
- *D* *antique furniture*

316 The horizontal timber which supports a common rafter midway between the eaves and the ridge is known as a

- *A* *hangar*
- *B* *soffit*
- *C* *purlin*
- *D* *binder*

317 The recommended moisture content for roof timbers is

- *A* *18%–20%*
- *B* *24%–26%*
- *C* *28%–30%*
- *D* *32%–34%*

318 Figure 50 shows loads being lifted by a two-leg sling at M and N

- *A* *M will lift a greater load than N*
- *B* *N will lift a greater load than M*
- *C* *M and N can lift the same maximum load*
- *D* *the maximum load that can be lifted depends upon the length of the chain*

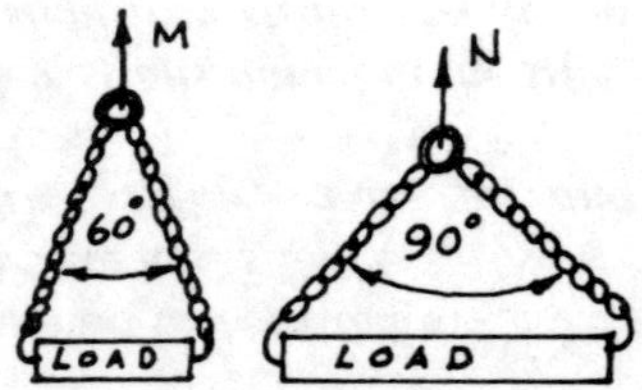

Figure 50

319 The minimum clearance between roofing timbers and the external face of a chimney flue is

A *20 mm*
B *38 mm*
C *60 mm*
D *75 mm*

320 The feet of jack rafters can be fixed to a valley rafter. An alternative is to fix them to

A *gutter board*
B *boarding joist*
C *barge board*
D *lay board*

321 Figure 51 shows an adult beetle which is about 25 mm long and has very long antennae (feelers). Its larva is commonly found in softwood roofing timbers. It is known as a

A *furniture beetle*
B *house longhorn beetle*
C *death watch beetle*
D *powder post beetle*

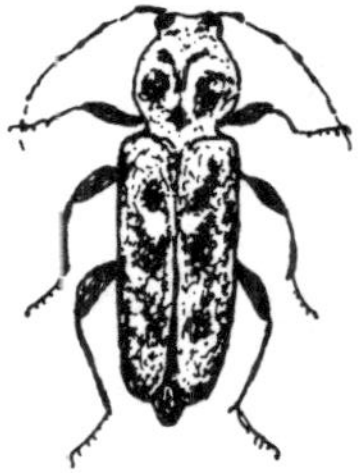
Figure 51

322 Upper floor joists, when over a certain span, require stiffening by

A *herringbone strutting*
B *solid bridging*
C *bracket struts*
D *noggings*

323 Which machine is used by a timber merchant to produce large quantities of softwood T. and G. floor-boards?

A *vertical spindle moulder*
B *panel planer*
C *four-sided planer and moulder*
D *combined planer and thicknesser*

324 A site method of applying preservatives to the ends only of upper floor joists which are to be built into brickwork, would be

- *A* *by use of bandage-type preservative*
- *B* *by hand brushing*
- *C* *by use of portable sprayer*
- *D* *by immersion in hot and cold tank*

325 A suitable hard-wearing timber for a block floor finish to an entrance hall is

- *A* *Yellow pine*
- *B* *Norway spruce*
- *C* *Japanese oak*
- *D* *Alaska pine*

326 Trestle scaffolds must not be used where the distance from the floor level to the working platform exceeds

- *A* *3·000 m*
- *B* *3·500 m*
- *C* *4·500 m*
- *D* *5·000 m*

327 Figure 52 is a line diagram of a domestic roof framing known as

- *A* *TRADA trussed rafter*
- *B* *'W' trussed rafter*
- *C* *bolted trussed rafter*
- *D* *framed trussed rafter*

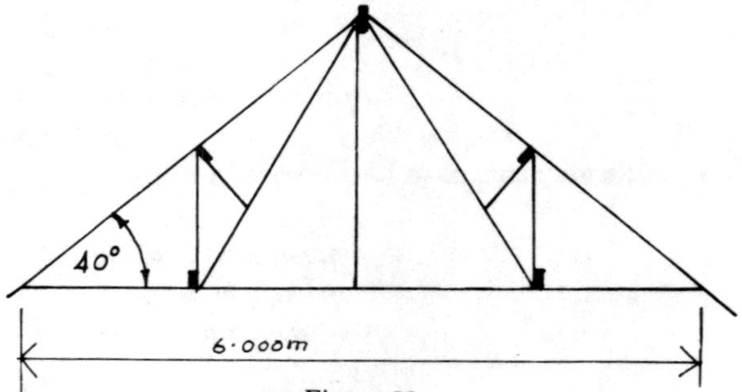

Figure 52

328 Most of the joints for the roof framing in Figure 52 are secured by

- *A* *bolts and split ring connectors*
- *B* *gangnail plates*
- *C* *ring shanked nails*
- *D* *bolts and tooth plate connectors*

329 The roof framing in Figure 52 would be spaced along the wall plate and ridge at centres of

- *A* *1·0 m*
- *B* *1·8 m*
- *C* *2·5 m*
- *D* *3·0 m*

330 The roof framing in Figure 52 would be made

- *A* *as a complete frame in the workshop and lifted into position on the site*
- *B* *in two halves in the workshop and joined together on the site*
- *C* *all members cut to length and the joints prepared in the workshop ready for assembly on the site*
- *D* *all members cut and joints made on the site*

331 The symbol in Figure 53 can be seen on certain power tools. This indicates that the tool

- *A* *is for operation by air pressure only*
- *B* *is double insulated*
- *C* *has been twice checked for safety*
- *D* *is of foreign manufacture*

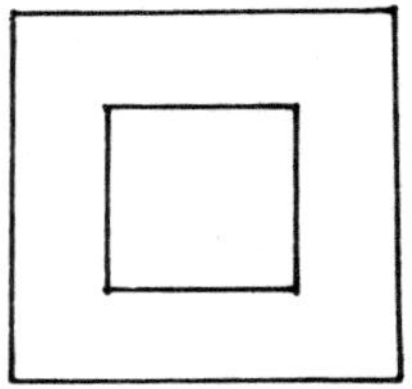

Figure 53

332 To reduce the risk of fire spreading from one dwelling to a neighbouring dwelling only two combustible materials are allowed to run over or through the separating wall. These are

A *roofing felt and battens*
B *purlins and binders*
C *ceiling joists and ridge*
D *wall plate and fascia*

333 The velocity ratio for the pulley system shown in Figure 54 is

A *2*
B *3*
C *4*
D *5*

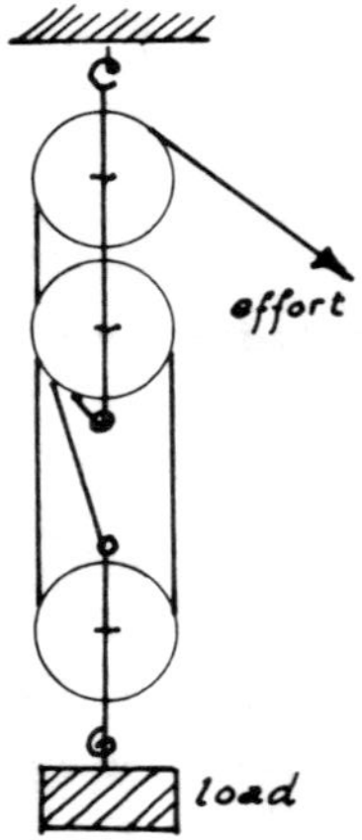

Figure 54

334 'Close laggings' are used in

A *centres for masonry arches*
B *forming a key for plaster ceilings*
C *centres for brick arches*
D *timbering to trench excavations*

335 Carcassing timbers are used in carpentry work for

A *making cupboards*
B *framing screen partitions*
C *making door casings*
D *framing roofs*

336 An oak newel post has a large number of minute holes packed with fine dust (frass); near the holes are some dead beetles about 5 mm long. This damage would be caused by

A *furniture beetle*
B *power post beetle*
C *wood wasp*
D *pin-hole borer*

337 Hardwood block flooring is supplied by the manufacturer with a low M.C. What precautions are taken on the site during its storage and fixing

A *delivered and fixed six months after the heating system has been operating*
B *delivered and fixed immediately after the cement screed has dried*
C *delivered and fixed when the heating system is completed and working*
D *delivered with the carcassing timbers, stored under cover and fixed before second fixings are commenced.*

338 Figure 55 shows temporary timbering supporting a load-bearing brick wall. It is known as

A *flying shore*
B *raking shore*
C *dead shore*
D *triangular shore*

339 The angle x in Figure 55 should be

A *60°*
B *75°*
C *85°*
D *105°*

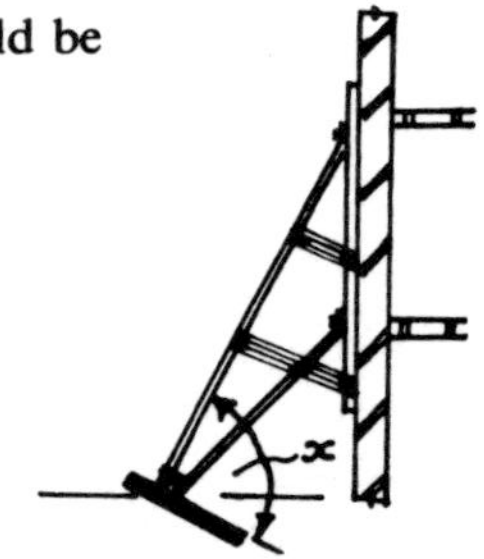

Figure 55

340 Site excavations must be suitably guarded when their depth exceeds

A *2·5 m*
B *2 m*
C *1·5 m*
D *1 m*

341 The maximum voltage shock an operator can receive from a faulty power tool connected to 110 V supply is

A *110 V*
B *55 V*
C *220 V*
D *75 V*

342 To ensure that trussed rafters can resist high wind pressures they should be fixed to the wall plate by

A *100 mm round wire nails*
B *100 mm oval wire nails*
C *100 mm screwdrive nails*
D *MS framing anchors*

343 In Figure 56 dimension L is known as the

A *effective length*
B *nett length*
C *span length*
D *mechanical length*

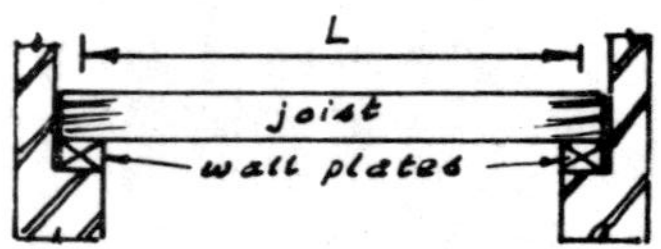

Figure 56

344 After stripping in-situ formwork timbers, an important safety precaution is to

A *clean faces of the timber*
B *grease faces of the formwork ready for re-use*
C *extract protruding nail*
D *burn all unwanted timber*

345 The minimum distance from the edge of masonry or concrete at which a cartridge-type fixing tool can be safely used is

A *25 mm*
B *38 mm*
C *63 mm*
D *100 mm*

346 Figure 57 shows an access ladder to a platform on a putlog scaffold. The distances a and b should be a minimum of

	a	b
A	*1·400 m*	*1·060 m*
B	*1·866 m*	*0·760 m*
C	*1·520 m*	*0·825 m*
D	*0·933 m*	*0·900 m*

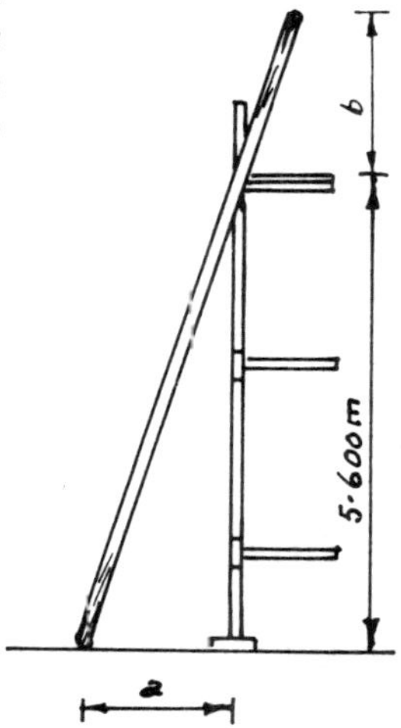

Figure 57

347 The mechanical advantage for the crowbar shown in Figure 58 is

A *780*
B *120*
C *60*
D *12*

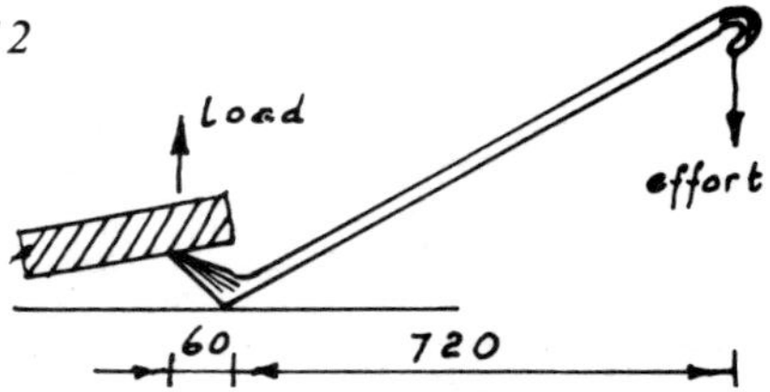

Figure 58

348 The volume of 50 metres run of 250 by 100 mm softwood timber is

A $1{\cdot}250\ m^3$
B $12{\cdot}500\ m^3$
C $62{\cdot}500\ m^3$
D $0{\cdot}612\ m^3$

349 The formwork for an in-situ concrete lintel of 2·000 m span will require

- *A* *sides and soffit to be of the same length*
- *B* *soffit to be longer than the sides*
- *C* *formwork to the sides only*
- *D* *sides to be longer than the soffit*

350 The ends of first floor joists, when supported by a brick party wall, should be fixed by

- *A* *building into the brickwork on to a softwood wall plate*
- *B* *building into the brickwork on to a wrought iron plate*
- *C* *building into the brickwork on to a cement mortar bed*
- *D* *nailing into a mild steel hanger suspended from the brickwork*

351 Figure 59 shows the end of a timber board, which is

- *A* *rift-sawn softwood*
- *B* *quarter-sawn hardwood*
- *C* *flat-sawn softwood*
- *D* *tangential-sawn hardwood*

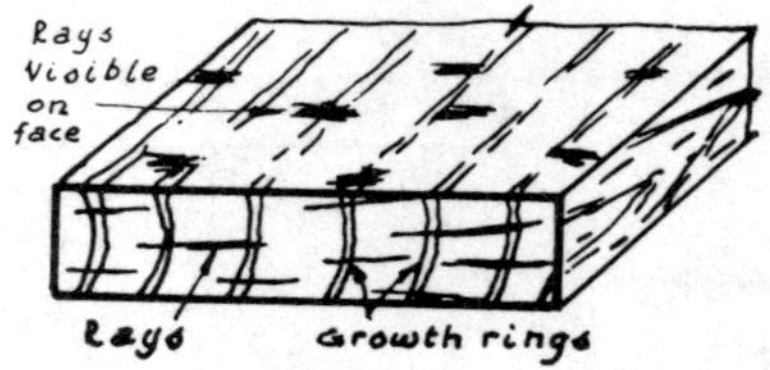

Figure 59

352 Figure 60 shows the pitch for a chain mortiser. The three pitches used for joinery mortising are

A *9 mm, 15 mm and 20 mm*
B *12 mm, 18 mm and 24 mm*
C *12 mm, 18 mm and 15 mm*
D *13 mm, 16 mm and 22 mm*

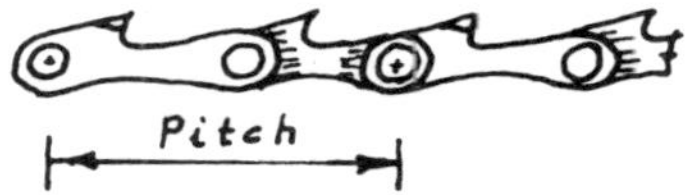

Figure 60

353 The term 'peripheral speed' is used for specifying the speed of

A *the circumference of a circular saw*
B *the circular motions per second of an orbital sander*
C *the revolutions per minute of a bandsaw wheel*
D *the spindle revolutions per second of an electric motor*

354 Adjacent panels in a panelled door are separated by a vertical member known as

A *mullion*
B *muntin*
C *stile*
D *jamb*

355 A step-down transformer is used for

A *increasing voltage*
B *suppressing electrical interference*
C *providing multiple socket outlets*
D *decreasing voltage*

356 Which of the following is most suitable for window cills?

A *Hemlock*
B *Iroko*
C *Beech*
D *Parana pine*

357 Specify, in correct sequence, the following routine safety points to be checked before using an electric power tool:

1 check the tool switch is in the off position
2 check the tool cutting edges are sharp
3 check the tool for correct voltage
4 check the plug for firm fixing to the cable
5 check the cable for visual faults
6 check personal clothing and eye protection
7 check the tool guards are in the correct position

A *3, 5, 4, 1, 7, 2, 6*
B *1, 7, 6, 4, 3, 5, 2*
C *3, 6, 4, 5, 2, 7, 1*
D *5, 2, 1, 6, 4, 3, 7*

358 Figure 61 shows a mortise chain on a guide bar. The amount of chain slack at x should be

A *4 mm*
B *6 mm*
C *8 mm*
D *10 mm*

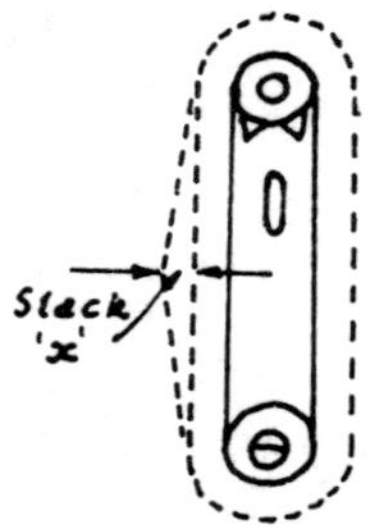

Figure 61

359 The sizes of electric cables for power tools are designated by a number, e.g. 40/0078. The first two figures indicate

A *number of strands in each wire*
B *gauge of the wire*
C *maximum length of the cable in metres*
D *diameter of the cable*

360 Which, in Figure 62, shows the correct 'set' for the teeth of a narrow bandsaw

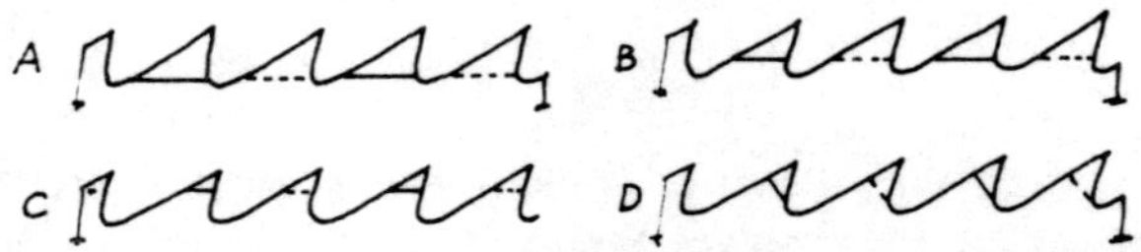

Figure 62

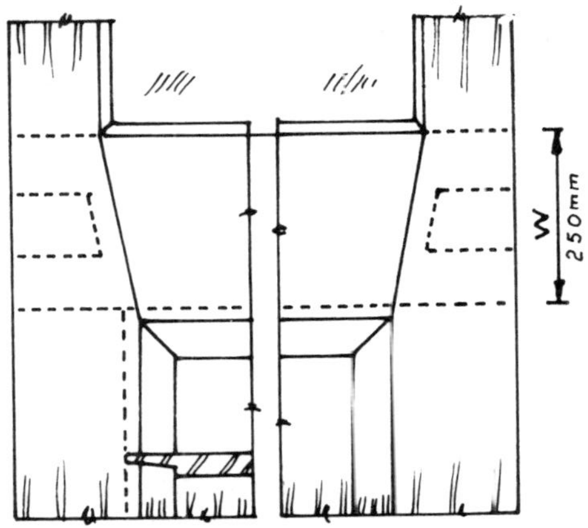

Figure 63

361 Figure 63 shows the part elevation of a panelled and glazed door. The purpose of haunching the centre of the tenons in the middle rail is to

A *provide waste material for cutting wedges*

B *prevent the buckling of wide tenons*

C *reduce the end grain showing on the edge of the door*

D *double the wedges*

362 The stiles in Figure 63 are referred to as being

A *diminished*

B *splayed*

C *tapered*

D *reduced*

363 A suitable width for each tenon in Figure 63 is

A $W \div 3$

B $W \div 4$

C $W \div 5$

D $W \div 6$

364 The internal glazing beads to the glass panel in Figure 63 are best secured by

- *A* *brassed panel pins*
- *B* *c/s mild steel screws*
- *C* *brass c/s screws and brass cups*
- *D* *contact adhesive*

365 Specifications for wood casement windows are detailed in British Standard No.

- *A* *459*
- *B* *1192*
- *C* *584*
- *D* *644*

366 At the end of a day's work the blade of a narrow bandsaw machine should

- *A* *be removed from the machine*
- *B* *have the tension eased off*
- *C* *be covered with a film of oil*
- *D* *have the saw guides lowered to table level*

367 In a multi-light casement window the vertical member separating adjacent lights is known as

- *A* *mullion*
- *B* *muntin*
- *C* *jamb*
- *D* *transome*

368 'Franking' or 'reverse haunching' is used in mortise and tenon joints for sashes because

- *A* *it makes machine production easier*
- *B* *it increases the width of the tenon*
- *C* *it increases the resistance to wedges*
- *D* *it reduces the amount of exposed end grain*

369 The meeting rails for vertical sliding sashes are made draughtproof by having their meeting faces

- *A* *splay-rebated*
- *B* *splayed only*
- *C* *square-rebated*
- *D* *hook-jointed*

370 Lipped or stormproof sashes produced to BS specifications are jointed by

A *mortise and tenon*
B *hardwood dowels*
C *alloy metal star dowels*
D *combing*

371 Isometric projection distorts the true value of angles in

A *plan only*
B *plan, front and side elevations*
C *front elevation only*
D *side elevation only*

372 Figure 64 shows the section through the jamb of a storey frame. The groove at the back is provided to

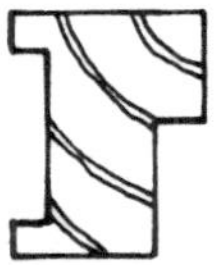
Figure 64

A *accommodate lightweight partition blocks*
B *bed the frame in cement mortar*
C *prevent the timber warping*
D *accommodate metal fixing anchors*

373 A suitable height for a domestic kitchen sink unit to BS 1195 is

A *850 mm*
B *900 mm*
C *950 mm*
D *1·000 m*

374 In Figure 65 the sides of the hexagon are equal to

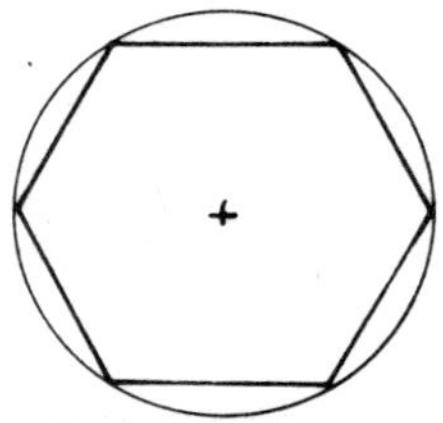
Figure 65

A $\frac{1}{6}$ *circumference of the circle*
B $\frac{1}{3}$ *diameter of the circle*
C *radius of the circle*
D *0·866 of the radius*

375 'Yorkshire lights' are casement windows with

- *A* *horizontal pivot sashes*
- *B* *side hung folding sashes*
- *C* *vertical pivot sashes*
- *D* *horizontal sliding sashes*

376 The pockets in a double hung vertical sliding sash window are removed to

- *A* *remove the sashes*
- *B* *give access to the weights*
- *C* *insert a mouse*
- *D* *give access for fixing to the wall*

377 Skeleton-framed flush doors for domestic cupboards should be covered both sides to

- *A* *prevent the frame from twisting*
- *B* *provide dust-free surfaces*
- *C* *give a solid appearance*
- *D* *economise in the paint or varnish finishing coat*

378 A transomed internal door casing runs from floor to ceiling. The glazed light above the transome is known as

- *A* *transome light*
- *B* *ceiling light*
- *C* *frieze light*
- *D* *borrowed light*

379 Figure 66 shows a sash stile without a horn, and meeting rail for vertical sliding sashes. A suitable joint would be

- *A* *open mortise and tenon*
- *B* *haunched mortise and tenon*
- *C* *dovetail*
- *D* *bridle*

Figure 66

380 The angle between adjacent sides of a regular pentagon is

A *108°*
B *54°*
C *72°*
D *36°*

381 In Figure 67 the diagonal of the square is equal to the circle

A *radius²*
B *diameter ÷ 4*
C *diameter ÷ 2*
D $\sqrt{}$*diameter*

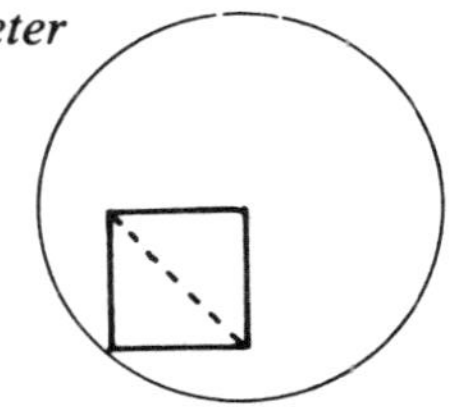

Figure 67

382 Figure 68 shows a 'tonk' strip used for supporting adjustable shelves. The slots in the metal strip are for

A *cast iron lugs*
B *steel rods*
C *cantilevered brackets*
D *metal studs*

Figure 68

383 When making a pair of drive entrance gates the braces should be positioned to resist

- *A* *compressive stresses*
- *B* *tensile stresses*
- *C* *bending stresses*
- *D* *sheer stresses*

384 Which dovetail joint is the most suitable to use between a drawer front and side?

- *A* *double lapped dovetail*
- *B* *through dovetail*
- *C* *lapped dovetail*
- *D* *dovetailed housing*

385 In Figure 69, p equals

- *A* *4·33*
- *B* *8·66*
- *C* *0·433*
- *D* *0·866*

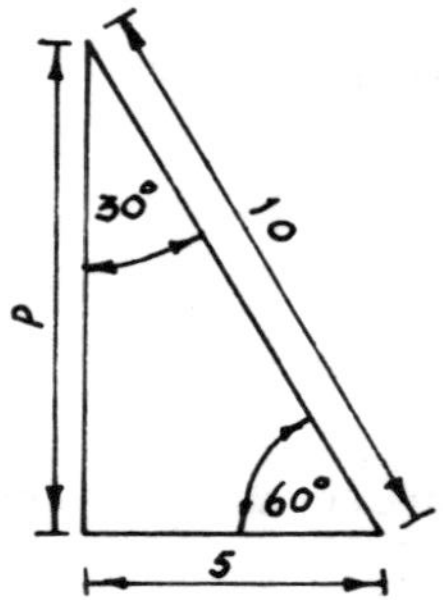

Figure 69

386 The movement of water between the edges of a casement and the rebate of a window frame is restricted by

- *A* *throatings*
- *B* *drips*
- *C* *anti-capillary grooves*
- *D* *weather-strip*

387 Figure 70 shows a joiner's tool used for

- *A* *striking circular curves*
- *B* *scribing profiles*
- *C* *setting-out elliptical curves*
- *D* *bradawls for equally-spaced screws*

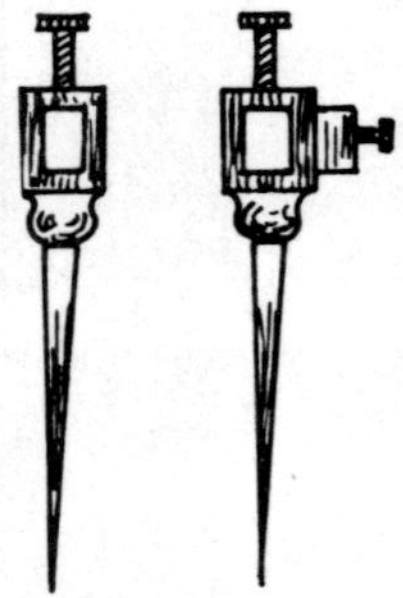
Figure 70

388 Which of the shapes in Figure 71 is an ogee moulding?

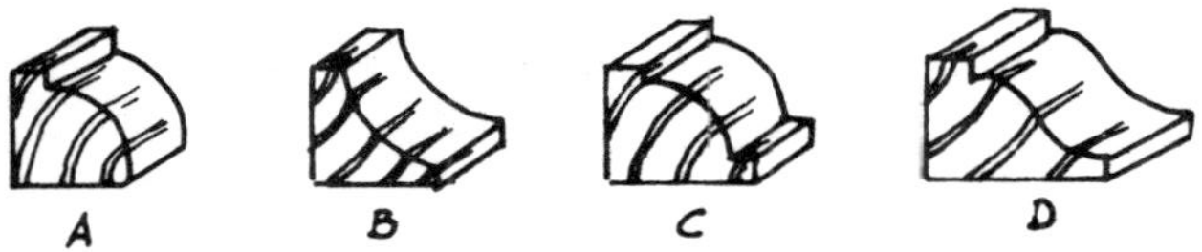

Figure 71

389 True teak is often specified for worktops to laboratory fitments because

A *it has a dark colour*
B *it is obtainable in narrow boards*
C *it has a high acid resistance*
D *it has a low density*

390 The setting time for resin adhesives can be accelerated by

A *increasing the temperature of the glue line*
B *decreasing the temperature of the glue line*
C *increasing the cramp pressure*
D *using a higher proportion of resin in the mix*

391 Diminished stiles in door construction are used to

A *provide larger glazed panels*
B *reduce the thickness of doors*
C *give a better appearance*
D *strengthen the stability of doors*

392 A flight of four treads to a pavilion verandah has a total rise of 750 mm. From the step formula $2R + G = 575$, G will be

A *225 mm*
B *250 mm*
C *275 mm*
D *300 mm*

393 Douglas fir plywood shows a pronounced tangential grain. This is produced by cutting the veneers

A *with a thin gauge bandsaw*
B *on a rotary lathe*
C *with a fixed slicing knife*
D *on a large diameter circular saw*

394 Figure 72 shows a fixing device known as

A *rawlanchor*
B *rawlnut*
C *spring toggle*
D *gravity toggle*

Figure 72

395 One of the advantages of using a 'Unique' spring balance for sliding sashes is

A *it can be used with solid frames*
B *it never needs replacing*
C *it is easier to fix than weights and cords*
D *it prevents the sashes from sticking*

396 In a small joiner's shop the following machines are available for producing a batch of casement windows:

1 straight line crosscut saw
2 ripsaw bench
3 spindle
4 combined surface and thickness planer
5 combined chain and chisel mortiser
6 single-end tenoner

The correct sequence for using these machines would be

A *2, 1, 4, 3, 5, 6*
B *1, 2, 4, 5, 6, 3*
C *1, 4, 2, 5, 6, 3*
D *4, 2, 1, 3, 6, 5*

397 The moisture content of timber can be calculated from the formula

$$\% \text{ M.C.} = \frac{\text{Wet weight} - \text{Dry weight}}{\text{Dry weight}} \times \frac{100}{1}$$

If the wet weight of a specimen is 23 gm and its dry weight is 20 gm, the % M.C. will be

A *0·15%*
B *1·5%*
C *11%*
D *15%*

398 Which adhesive should be used with care in confined spaces due to its inflammable fumes?

A *casein cement*
B *contact adhesive*
C *blood albumen*
D *polyvynal acetate glue*

399 The formula $2\pi rh$ is used to calculate the curved surface area of a right circular cylinder. When $r = 700$ mm and $h = 1{\cdot}500$ m the area will be

A *6·600* m^2
B *660 000* mm^2
C *4·620* m^2
D *46·20* m^2

400 The best finishing treatment for a builder's rung-ladder is

A *pressure impregnation with creosote preservative*
B *two undercoats and one gloss coat of red paint*
C *two coats of clear external-grade varnish*
D *no treatment, leave the timber in the white*

401 Which of the threads in Figure 73 is a twin fast wood screw?

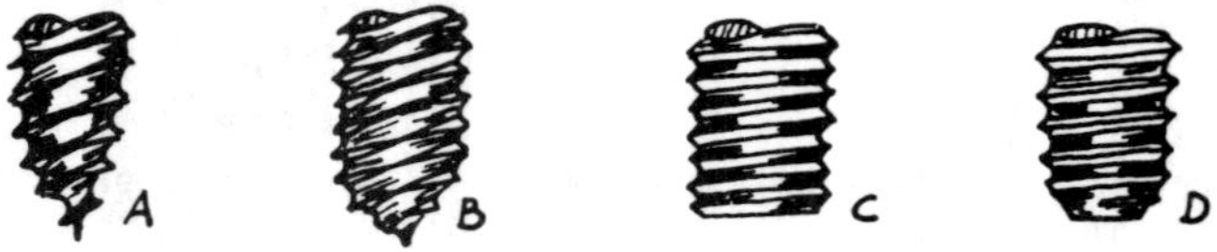

Figure 73

402 A large picture window is fitted with a single pivot sash opening horizontally. A suitable fastener for this type of sash is

A *spiral spring rods*
B *sash screws*
C *cockspur lever*
D *single lever espagnolette bolts*

403 When working with asbestos, epoxy resins or mould oils, particular precautions should be taken against

A *low temperatures*
B *health hazards*
C *blunting of edge tools*
D *low flash-point*

404 The maximum current through a portable electric power tool is 4 amperes. Which cartridge fuse, of the following standard current ratings, should be used to protect the tool?

A *2 amperes*
B *5 amperes*
C *10 amperes*
D *13 amperes*

405 One advantage of pressure impregnation of preservatives is

A *deep penetration into the timber section*
B *provides a uniform colour*
C *uses the minimum amount of preservative*
D *stabilises the M.C. of the timber*

406 Progressive kilns are particularly useful for seasoning timber

A *of long lengths and large section*
B *which require a long period of drying*
C *in large quantities of the same species*
D *of mixed species and various sectional sizes*

407 Which of the graphical symbols in Figure 74 for the end grain of unwrot timber complies with BS 1192?

Figure 74

408 'Wash board' or timber collapse is a defect caused by

A *a wet rot fungus*
B *excessive weathering on an exposed site*
C *excessive vacuum in a preservative pressure tank*
D *over-rapid kiln drying*

409 Structural softwoods are purchased from a timber merchant at a unit price

A *per metre cube*
B *per metre square*
C *per metre lineal*
D *per 100 metre lineal*

410 The SI unit for expressing the load capacity of a roof surface is

A N/mm^2
B kgF/m^2
C kgF/mm^2
D N/m^2

411 Which of the right angled triangles in Figure 75 would give the true length, plumb and seat bevels for a hip rafter?

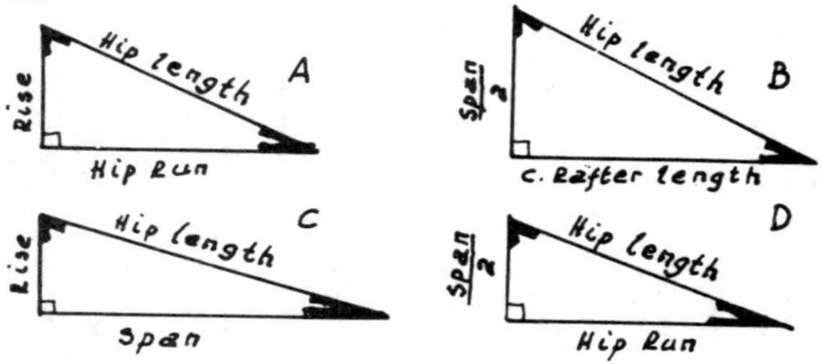

Figure 75

412 Toothed plate connectors and bolts are used in carpentry framing joints to

A *provide an easy way of dismantling the frames*
B *allow for expansion and shrinkage of the joint*
C *spread the load over a larger area*
D *reduce the production cost of making the frames*

413 Hardwood floor blocks can be obtained with a cork strip inserted as shown in Figure 76. These are used to

A *allow for expansion and contraction of the floor finish.*
B *provide greater resilience for a dance floor*
C *absorb excessive moisture in high humidity*
D *enable the last row of blocks to be laid*

Figure 76

414 Which nail would be used for nailing hardwood floorboards prepared for secret fixing?

A *ring shank*
B *cut floor brad*
C *plain wire nail*
D *round lost head*

415 The ends of the herringbone strutting shown in Figure 77 are cut to clear the joist edges by approximately 12 mm. The reason for this is to

- *A* *prevent disturbance of the floorboards or ceiling due to shrinkage of the joists*
- *B* *provide a clearance for the fixing nails*
- *C* *increase the angle of thrust between the strut and joist*
- *D* *allow for variations in the lengths of the struts*

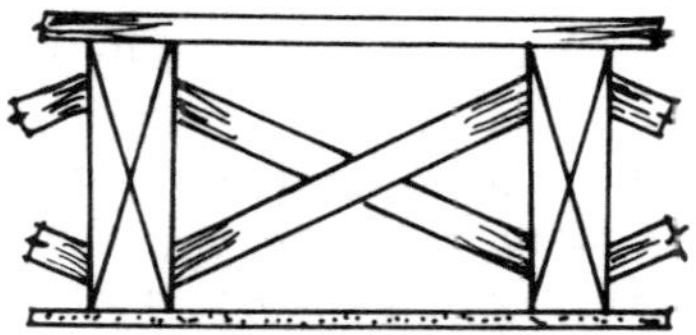

Figure 77

416 Figure 78 shows, in line diagram, the plan of roof members. The members x fixed at the valley and hip are known as

- *A* *jack rafters*
- *B* *valley rafters*
- *C* *hip rafters*
- *D* *cripple rafters*

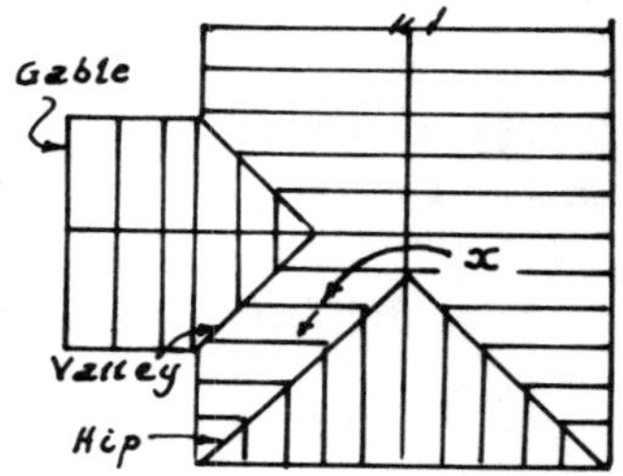

Figure 78

417 The newton is the force which, applied to a mass of 1 kg, will

- *A* *move it a distance of 1 m*
- *B* *give it an acceleration of 1 m per sec per sec*
- *C* *move it a distance of 1 m every second*
- *D* *give it an acceleration of 1 m every 2 seconds*

418 The board in Figure 79 would be used for

- *A* *secret-nailed hardwood floorboard*
- *B* *vertical cladding to a gable wall*
- *C* *projecting cill drip*
- *D* *hardwood window board*

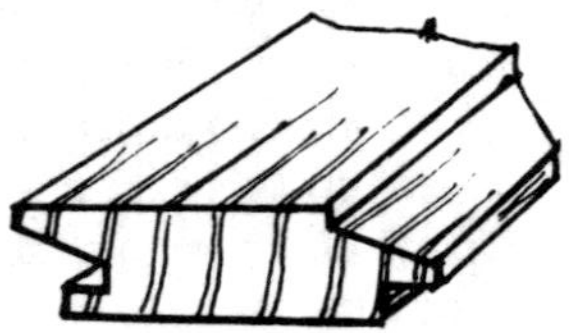

Figure 79

419 Which timber connector will carry the highest permissible load per fixing?

- *A* *framing anchor*
- *B* *single-side tooth plate*
- *C* *split ring*
- *D* *double-side tooth plate*

420 Figure 80 shows adjustable steel props supporting a timber beam with a central load. This type of loading is known as

- *A* *distributive load*
- *B* *point load*
- *C* *eccentric load*
- *D* *safe load*

421 If the timber beam in Figure 80 is loaded to cause bending, the effect on the bottom edge will be to cause excessive

- *A* *tension*
- *B* *compression*
- *C* *shear*
- *D* *cleavage*

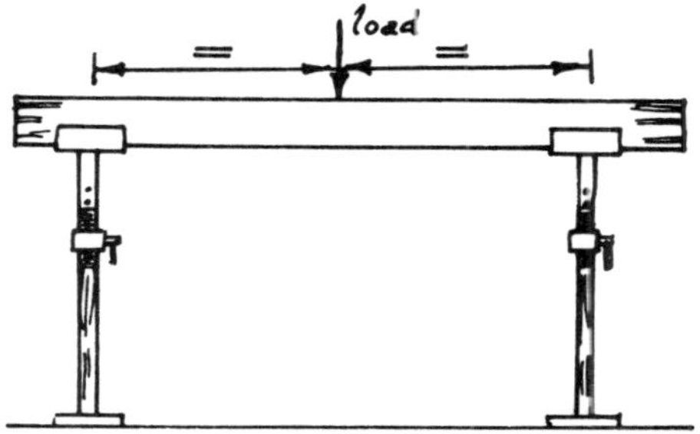

Figure 80

422 If the load in Figure 80 is 1000 kg, each steel prop will support

- *A* *125 kg*
- *B* *250 kg*
- *C* *500 kg*
- *D* *1000 kg*

423 The true shape of the plane x–x cutting through the right circular cone in Figure 81 will be

- *A* *an ellipse*
- *B* *a triangle*
- *C* *a parabola*
- *D* *a hyperbola*

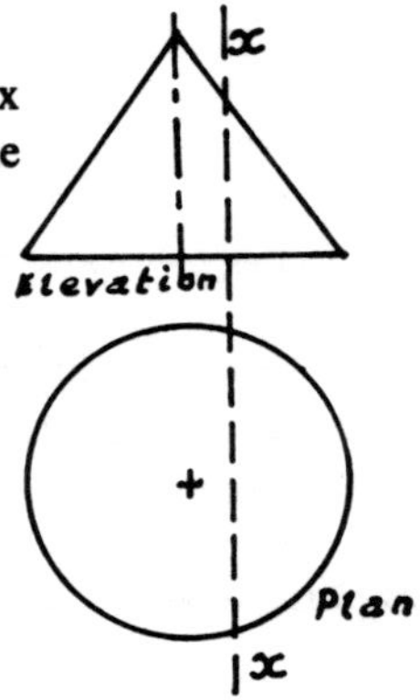

Figure 81

424 If $V = \frac{1}{3}\pi r^2 h$ then $r^2 =$

A $\dfrac{V}{\frac{1}{3}\pi h}$

B $\frac{1}{3}V \div \pi h$

C $V + \frac{1}{3}\pi h$

D $\sqrt{\dfrac{V}{\frac{1}{3}\pi h}}$

425 Figure 82 shows a diagram of the

A *first order of levers*

B *second order of levers*

C *third order of levers*

D *fourth order of levers*

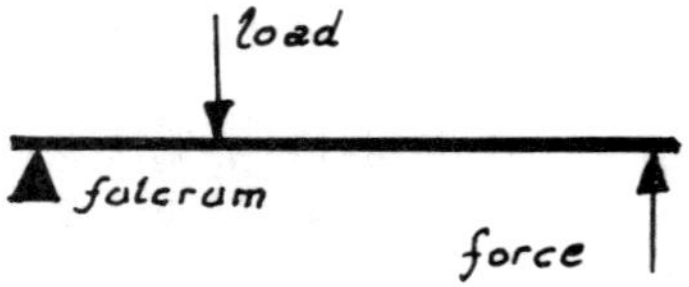

Figure 82

426 Large metal dogs are used

A *as a metal fastening for heavy temporary timbering*

B *as bearing plates for heavy timber beams*

C *for binding together the base of raking shores*

D *as metal folding wedges in dead shores*

427 A flat roof of 5 m^2 supports a force of 200N. The pressure on the roof will be

A *1000* N/m^2

B *200* N/m^2

C *100* N/m^2

D *40* N/m^2

428 Carpentry timbers are stress-graded to determine their

A *thermal conductivity*

B *strength properties*

C *percentage of porosity*

D *fire resistance*

429 The arrangement of cells in beech is described as
- *A* *diffuse porous*
- *B* *ring porous*
- *C* *semi-ring porous*
- *D* *terminal*

430 Which painted surface will reflect the greatest amount of heat?
- *A* *gloss white*
- *B* *gloss black*
- *C* *matt white*
- *D* *matt black*

431 Dimension x in Figure 83 will be
- *A* *3 m*
- *B* *150 mm*
- *C* *300 mm*
- *D* *1500 mm*

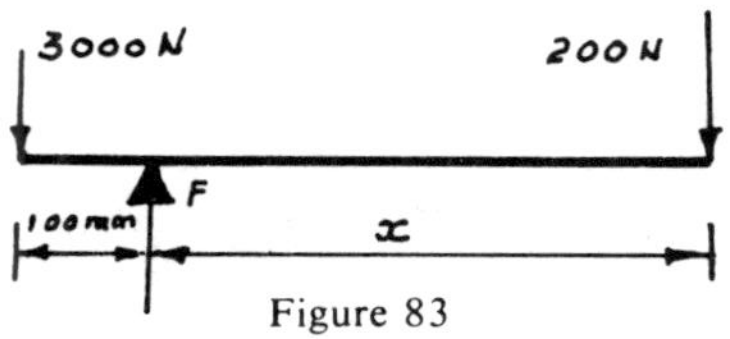

Figure 83

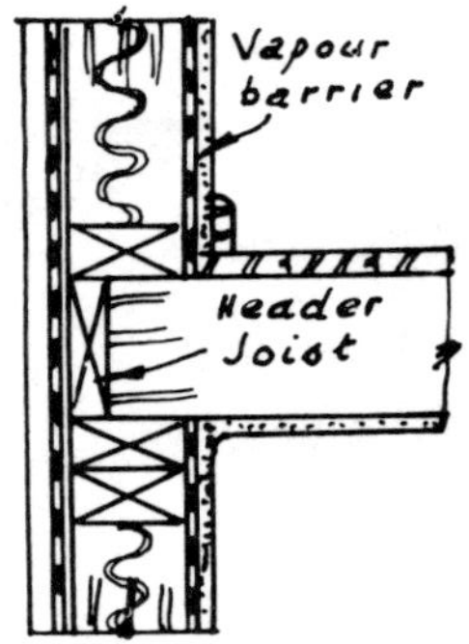

Figure 84

432 Figure 84 is a sketch detail between the external wall and upper floor of a timber framed construction known as
- *A* *balloon frame*
- *B* *platform frame*
- *C* *stud frame*
- *D* *post and beam frame*

433 A suitable material for the vapour barrier in Figure 84 would be

A *polythene sheeting*
B *fibre-glass quilt*
C *breather-type kraft paper*
D *bituminous sarking felt*

434 The function of the header joist in Figure 84 is to

A *serve as a binder for the floor joists*
B *transmit the upper floor loads to the lower floor*
C *provide a fire check between floors*
D *provide a fixing for the wall cladding*

435 A timber flat roof, with level joists, has a slight fall formed by fixing on top of the joists

A *firring pieces*
B *tilting fillets*
C *weathering fillets*
D *counter battens*

436 The object of discontinuous techniques of construction between walls and floors is

A *to prevent capillary attraction between surfaces*
B *to comply with fire-resistance regulations*
C *necessary with system building*
D *to improve sound insulation from impact noises*

437 Figure 85 is a sketch of a timber roof used for large clear spans. It is known as

A *crescent truss*
B *bowstring truss*
C *warren truss*
D *scissor truss*

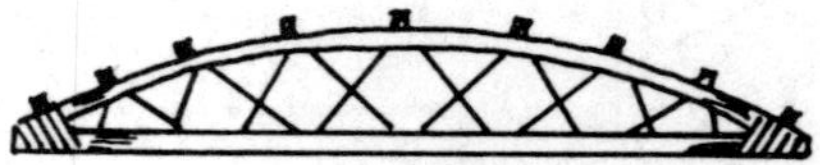

Figure 85

438 If the roof truss in Figure 85 were covered with plywood sheets size 1·220 × 2·440 m, a suitable spacing for the purlins would be

A *610 mm*
B *760 mm*
C *915 mm*
D *1220 mm*

439 A minimum ratio of rise to span for the roof in Figure 85 is

A *1 : 8*
B *1 : 10*
C *1 : 15*
D *1 : 20*

440 Figure 86 shows the plan of two inclined roof surfaces intersecting at a valley. The dihedral angle will be

A *k*
B *l*
C *m*
D *n*

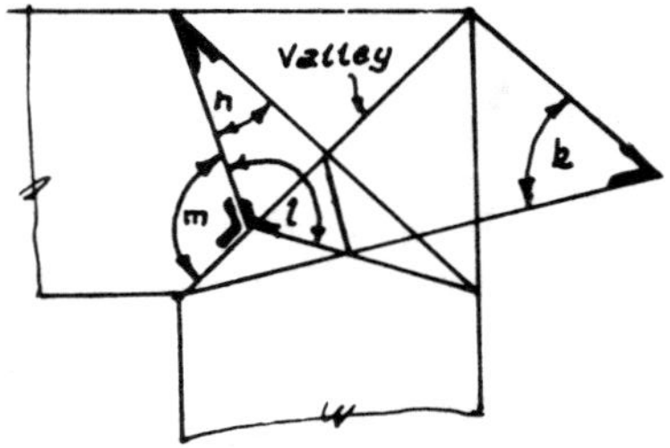

Figure 86

441 Joinery ironmongery can be specified as 'B.M.A. finish'. This means

A *brassed metal antique*
B *bright metal antique*
C *bronze metal antique*
D *buffed metal antique*

442 A suitable temporary support for the bridging joists to an upper floor while a faulty binder is replaced, would be

A *dead shores*
B *raking shores*
C *underpinning*
D *framed centres*

443 The hand tool shown in Figure 87 is used for

A *tightening nuts on handrail bolts*
B *lifting existing floorboards*
C *chopping out stiles for pulley wheels*
D *chopping out lock rails for mortise locks*

Figure 87

444 The top end of a raking shore is secured to the wallplate by

A *needle and cleat*
B *150 mm round wire nails*
C *bridle mortise and tenon*
D *bolts with timber connectors*

445 Bituminous felt flat roofs are covered with a final layer of stone chippings to

A *provide an attractive finish to the roof surface*
B *prevent the bitumen cracking in very cold weather*
C *prevent the bitumen melting in very hot weather*
D *provide a sound insulation barrier*

446 In Figure 88 the member x is known as a

A *poling board*
B *waler*
C *puncheon*
D *sheet boarding*

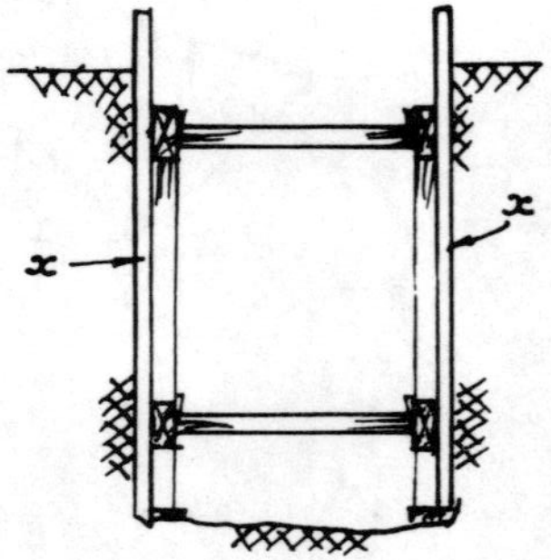

Figure 88

447 Dissimilar metals, when brought into contact in the presence of moisture, can cause a reaction known as

- *A* *oxidising*
- *B* *temperising*
- *C* *electrolytic*
- *D* *tarnishing*

448 Temporary timbering in shoring, planking and strutting, etc., should be inspected by a competent person

- *A* *once a fortnight*
- *B* *before start of work each day*
- *C* *every Monday morning*
- *D* *twice a week*

449 Steel beams, when used in house building, have to be encased to

- *A* *increase their fire resistance*
- *B* *improve their aesthetic appearance*
- *C* *prevent rusting*
- *D* *increase their compressive stress*

450 Which timber can cause corrosion to iron fittings?

- *A* *Parana pine*
- *B* *Western red cedar*
- *C* *Norway spruce*
- *D* *European larch*

451 Circular saws are provided with felt packing to

- *A* *guide and steady the saw blade*
- *B* *lubricate the saw blade*
- *C* *deaden the high-pitched noise of the saw*
- *D* *prevent the saw blade overheating*

452 A draw-bore tool is used for

- *A* *rough shaping of thin boards*
- *B* *easing the guides in a drawer opening*
- *C* *boring shallow recesses for flush drawer palls*
- *D* *tightening the shoulders of a dowelled mortise and tenon joint*

453 A false fence is used to

A *provide a snow fence for an exposed highway*
B *form temporary boarding to a building site*
C *bridge the cutter gap on a spindle moulder*
D *provide mobile barriers around excavations*

454 With most timbers the minimum shrinkage will occur

A *along the cambium layer*
B *on the radial plane*
C *across the tangential surface*
D *in the sapwood region*

455 A chain mortiser is provided with a chipbreaker to

A *guide the chippings into the extractor hood*
B *prevent the timber from splintering*
C *provide a guard for the operator's hands*
D *serve as a depth stop*

456 A ring fence is used on

A *router*
B *lathe*
C *spindle moulder*
D *bandsaw*

457 'Jointing' on a panel planing machine is done to

A *ensure all cutting edges are in the same cutting circle*
B *re-grind the cutter in the cutter block*
C *ensure that the cutters are dynamically balanced*
D *check that opposite cutters are set parallel*

458 Plywood sheets stamped B/BB would signify

A *type of adhesive used for bonding the plywood*
B *country of origin where the plywood was manufactured*
C *manufacturer's code for date of production*
D *quality of the plywood faces*

459 A short-circuit in an electric power tool is a path of

A *very high resistance which causes a very high current to flow*

B *very low resistance which causes a very low current to flow*

C *very low resistance which causes a very large current to flow*

D *very high resistance which causes a very low current to flow*

460 Figure 89 shows a section through the transom for a stormproof casement window. The upper sash is top-hung and the lower sash is side-hung, to open outwards. A fault with this detail is

A *the gap between sashes and transom is too large*

B *lack of a drip to the upper sash*

C *the glazing rebates are on the wrong side*

D *there is no fault–the detail is acceptable*

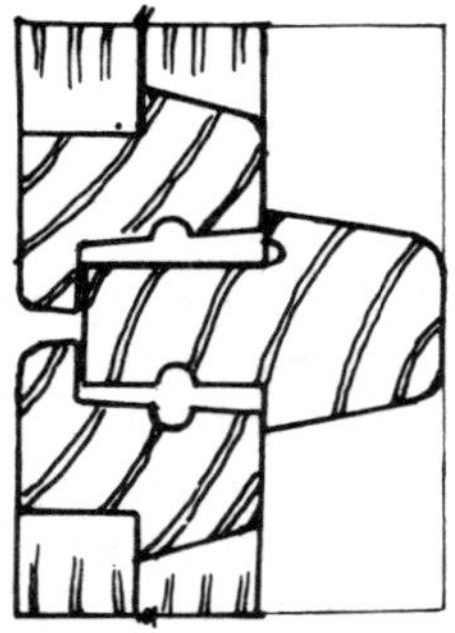

Figure 89

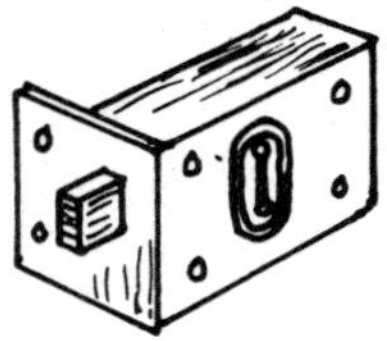

Figure 90

461 Figure 90 illustrates a dead

A *rim latch*

B *mortise lock*

C *rim lock*

D *cylinder lock*

462 An electric meter used for measuring the moisture content of timber operates upon the principle of

- *A* *the electrical conductivity of timber*
- *B* *electrolytic reaction between wood and metal*
- *C* *radiation of electrical energy*
- *D* *convection currents of electrical heat*

463 The aperture dimensions for a letter plate to BS 2911 are

- *A* *150 × 35 mm*
- *B* *200 × 35 mm*
- *C* *150 × 44 mm*
- *D* *200 × 44 mm*

464 The nett area of 6 mm plywood required for both faces of a flush door, size 1·950 × 0·750 m, is

- *A* $2{\cdot}925\ m^2$
- *B* $1{\cdot}462\ m^2$
- *C* $3{\cdot}900\ m^2$
- *D* $4{\cdot}387\ m^2$

465 A quantity of wedges for mortise and tenon joints is best made by

- *A* *using a purpose-made jig on a circular saw*
- *B* *ripsawing a bevelled length of timber*
- *C* *purchasing ready-made from a timber merchant*
- *D* *sawing any convenient waste timber on the band-saw*

466 Figure 91 shows a door detail from BS 459 Part 3, designed to provide

- *A* *sixty minutes' fire check*
- *B* *sound insulation against impact noises*
- *C* *thirty minutes' fire check*
- *D* *sound insulation against airborne noises*

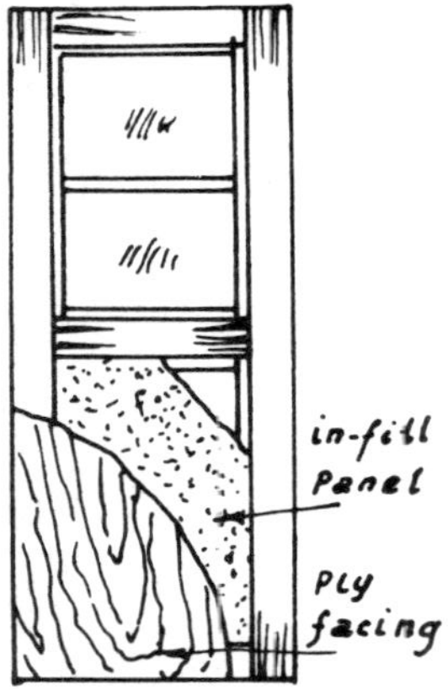

Figure 91

467 The in-fill panel shown in Figure 91 is made from

A *chipboard*
B *plasterboard*
C *fibreglass*
D *polystyrene*

468 The minimum and maximum heights for fixing BS 2911 letter plates are

	Minimum	Maximum
A	*750 mm*	*1·440 m*
B	*600 mm*	*1·440 m*
C	*750 mm*	*1·250 m*
D	*600 mm*	*1·250 m*

469 The term 'door stock' refers to

A *standard design doors manufactured for stock supply*
B *timber specifically graded for door stiles and rails*
C *diminished stiles used in half-glazed doors*
D *grooved and ovolo moulded timber supplied in random lengths for making 'one-off' doors*

470 A suitable hinge for self-closing lightweight doors to a suite of w.c. cubicles would be

A *helical hinge*
B *rising butt*
C *parliament hinge*
D *dolphin hinge*

471 The pilot hole for a No. 10 brass wood screw in hardwood requires a drill size

A *2·5 mm*
B *3·0 mm*
C *3·5 mm*
D *4·0 mm*

472 A preservative for treating softwood food-storage cabinets should be

A *an organic solvent*
B *tar oil derivative*
C *spirit based*
D *water-borne salts*

473 Lipped stormproof sashes are hung with

A *extension hinges*
B *cranked hinges*
C *strap hinges*
D *helical hinges*

474 A hammer-headed key joint is used when

A *making circular window frames*
B *fixing hammer shafts*
C *butt-jointing continuous handrails*
D *end-jointing laboratory worktops*

475 Figure 92 shows the plan and elevation of a skirting to an inclined ramp. The true section for the level skirting at x will be

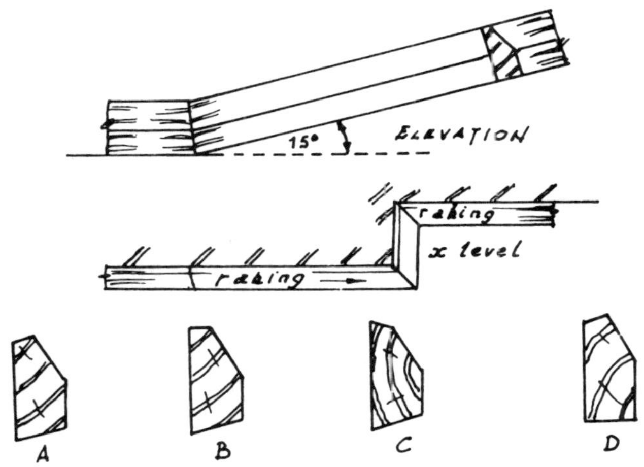

Figure 92

476 Heat from a hot water radiator is transferred to its surrounding atmosphere by

A *conduction*
B *radiation*
C *convection*
D *transference*

477 Figure 93 is a sketch of a cutter block used on a spindle moulder; it is known as

A *English head*
B *French head*
C *circular safety block*
D *whitehill block*

Figure 93

478 A particular advantage of the cutters used in the cutter block in Figure 93 is

A *a cutter projection scale is not required when grinding them to shape*
B *their cutting action has a positive cutting angle*
C *they provide a high degree of safety*
D *heavy deep cuts can be machined in one operation*

479 The metal foil backing on some plasterboards is intended to serve as a

A *sound insulator*
B *moisture barrier*
C *thermal insulator*
D *vermin barrier*

480 One way of preventing excessively wide cuttermarks when machine-planing joinery timbers is to

A *increase the feed speed*
B *decrease the feed speed*
C *increase the pressure on the feed rollers*
D *resharpen the cutters*

481 The traditional reason for framing wall panelling is

A *to permit the use of thin solid panels*
B *to enable secret fixing to the grounds*
C *to reduce the weight of the timber*
D *to economise with expensive hardwoods*

482 2·400 × 1·200 m reeded hardboard sheets being used for wall panelling should have their timber grounds spaced at x centres and be nailed at y centres

	Centres x	Centres y
A	*300–400 mm*	*100–150 mm*
B	*400–500 mm*	*100–200 mm*
C	*400–600 mm*	*75–175 mm*
D	*200–600 mm*	*75–200 mm*

483 Spandrel panelling is used to enclose a space

- *A* *under a sink*
- *B* *around a bulkhead*
- *C* *under a staircase*
- *D* *to a recess beside a chimney breast*

484 The physical change of water from a vapour to a liquid is known as

- *A* *vaporising*
- *B* *condensation*
- *C* *solidification*
- *D* *saturation*

485 Figure 94 shows the intersection of splayed linings to a square door opening. The mitre bevel will be

- *A* *E*
- *B* *F*
- *C* *G*
- *D* *H*

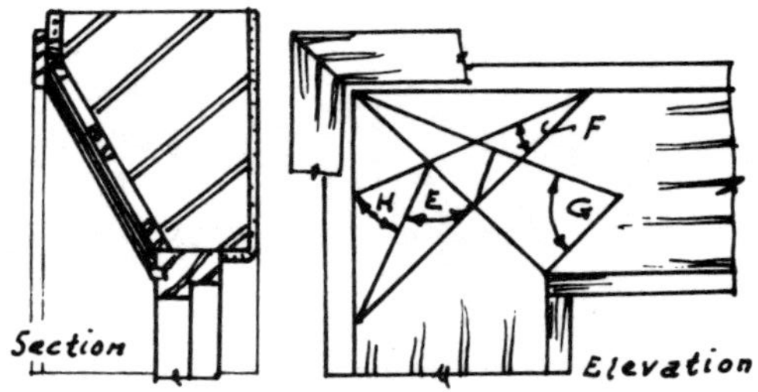

Figure 94

486 'Second Seasoning' is

- *A* *re-seasoning timber which has been dried too rapidly*
- *B* *kiln-seasoning timber which has been previously air-seasoned*
- *C* *conditioning of framed joinery after machining and before final fitting and assembly*
- *D* *re-seasoning imported timbers after a long sea journey*

487 Figure 95 shows a BS 1285 detail section through the head of a casement window designed for

A *double-glazed sealed units*
B *metal casements*
C *direct glazing*
D *double-lipped sashes*

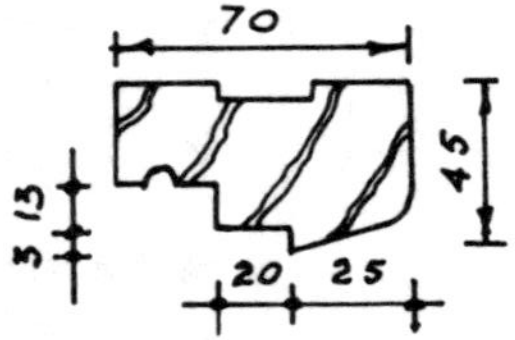

Figure 95

488 Some hardwood window boards have an interlocked grain; a suitable tool for finishing them to receive a clear finish is

A *block plane*
B *flat cabinet rasp*
C *disc sander*
D *hand scraper*

489 Figure 96 shows the BS 1192 graphical symbol for

A *joists to a flat roof*
B *flight of stairs*
C *studs to a partition*
D *boards to a fence*

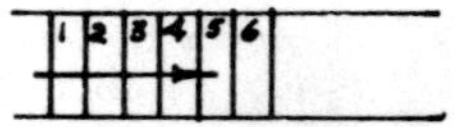

Figure 96

490 The front edge of a landing step is finished with a

A *lining*
B *capping*
C *nosing*
D *margin*

491 A floor-to-floor storey rod is set-out to provide twelve treads in a height of 2·34 m. Using the formula $2R + G = 600$, the 'going' per step will be

A *240 mm*
B *250 mm*
C *260 mm*
D *270 mm*

492 The contact surfaces of formwork are treated with mould oil to

A *accelerate the initial set period for the placed concrete*
B *ensure a glazed waterproof finish to the set concrete*
C *prevent the water in the wet mix soaking into the formwork*
D *prevent the hardened concrete adhering to the formwork surfaces*

493 The minimum headroom for a staircase, measured at 90 degrees from the pitch line, is

A *1·650 m*
B *1·500 m*
C *1·800 m*
D *1·950 m*

494 A concrete mix of 1 : 2 : 4 signifies

	Water	*Cement*	*Sand*	*Aggregate*
A	—	2	1	4
B	1	4	1	2
C	—	1	2	4
D	2	1	4	1

495 A floor-to-floor dimension of 2·625 m is divided by 15 to give a rise of 175 mm per step. How many treads will be required in a straight flight?

A *13*
B *14*
C *15*
D *16*

496 The maximum pitch for private stairways is

A *32°*
B *38°*
C *42°*
D *48°*

497 Building Regulations require the maximum rise and minimum go for a private stair to be

	Maximum rise	Minimum going (per step)
A	*175 mm*	*255 mm*
B	*150 mm*	*245 mm*
C	*188 mm*	*230 mm*
D	*220 mm*	*220 mm*

498 Building Regulations require a handrail to be provided to a flight of steps when

A *they are exposed to all weather conditions*
B *the total rise exceeds 600 mm*
C *they are to be used by the general public*
D *they are open treads, without risers*

499 A suitable site method for securing the mortise and tenon joint of a stair string to a newel post is

A *draw-bored dowels*
B *fox wedges*
C *handrail bolts*
D *corrugated fasteners*

500 In softwood timber (conifers) the majority of the cells are known as

A *vessels*
B *fibres*
C *tracheids*
D *soft tissue*

1	B	26	C	51	C	76	C	101	B	126	A	151	B	176	B	201	A	226	A
2	C	27	D	52	A	77	A	102	C	127	B	152	A	177	C	202	D	227	C
3	D	28	B	53	C	78	B	103	B	128	D	153	C	178	D	203	A	228	A
4	B	29	A	54	B	79	D	104	A	129	B	154	B	179	A	204	C	229	D
5	A	30	D	55	D	80	A	105	D	130	C	155	D	180	D	205	A	230	B
6	D	31	B	56	D	81	C	106	A	131	A	156	B	181	C	206	D	231	A
7	C	32	A	57	D	82	A	107	C	132	C	157	A	182	A	207	A	232	C
8	B	33	D	58	A	83	D	108	D	133	D	158	C	183	B	208	C	233	B
9	C	34	C	59	B	84	A	109	B	134	B	159	D	184	D	209	A	234	C
10	A	35	A	60	C	85	C	110	A	135	A	160	C	185	B	210	B	235	A
11	B	36	B	61	A	86	A	111	C	136	A	161	C	186	B	211	B	236	D
12	A	37	D	62	A	87	B	112	B	137	C	162	B	187	A	212	A	237	A
13	C	38	A	63	D	88	C	113	D	138	B	163	A	188	D	213	C	238	B
14	D	39	C	64	C	89	D	114	B	139	C	164	B	189	B	214	A	239	A
15	B	40	A	65	D	90	B	115	D	140	A	165	A	190	A	215	D	240	B
16	A	41	A	66	B	91	A	116	B	141	A	166	C	191	B	216	B	241	C
17	D	42	A	67	A	92	B	117	C	142	D	167	A	192	D	217	A	242	D
18	B	43	B	68	C	93	D	118	A	143	B	168	A	193	C	218	D	243	D
19	C	44	C	69	D	94	C	119	B	144	C	169	C	194	B	219	B	244	A
20	D	45	A	70	B	95	B	120	D	145	D	170	D	195	D	220	C	245	B
21	A	46	B	71	B	96	B	121	B	146	B	171	A	196	A	221	D	246	B
22	C	47	A	72	A	97	A	122	C	147	A	172	B	197	C	222	A	247	C
23	B	48	D	73	C	98	D	123	A	148	C	173	D	198	A	223	B	248	C
24	A	49	C	74	D	99	C	124	D	149	A	174	C	199	A	224	B	249	A
25	D	50	B	75	B	100	A	125	D	150	B	175	D	200	D	225	D	250	B

251	B	276	B	301	B	326	C	351	B	376	B	401	B	426	A	451	A	476	C
252	A	277	C	302	A	327	A	352	D	377	A	402	D	427	D	452	D	477	B
253	D	278	C	303	B	328	D	353	A	378	D	403	B	428	B	453	C	478	A
254	B	279	C	304	D	329	B	354	B	379	C	404	B	429	A	454	B	479	C
255	C	280	B	305	B	330	B	355	D	380	C	405	A	430	A	455	B	480	B
256	A	281	B	306	A	331	B	356	B	381	C	406	C	431	D	456	C	481	A
257	B	282	C	307	D	332	A	357	A	382	D	407	A	432	B	457	A	482	A
258	B	283	A	308	A	333	B	358	B	383	A	408	D	433	A	458	D	483	C
259	C	284	B	309	B	334	C	359	A	384	C	409	A	434	C	459	A	484	B
260	A	285	C	310	A	335	D	360	C	385	B	410	D	435	A	460	B	485	A
261	B	286	C	311	B	336	A	361	B	386	C	411	A	436	D	461	C	486	C
262	A	287	B	312	A	337	C	362	A	387	A	412	C	437	B	462	A	487	B
263	D	288	A	313	C	338	B	363	B	388	D	413	A	438	A	463	D	488	D
264	A	289	B	314	D	339	C	364	C	389	C	414	D	439	B	464	A	489	B
265	C	290	C	315	A	340	B	365	D	390	A	415	A	440	B	465	A	490	C
266	A	291	A	316	C	341	B	366	B	391	A	416	D	441	C	466	C	491	A
267	C	292	B	317	A	342	D	367	A	392	C	417	B	442	A	467	B	492	D
268	B	293	C	318	A	343	A	368	C	393	B	418	A	443	D	468	A	493	B
269	A	294	A	319	B	344	C	369	A	394	D	419	C	444	A	469	B	494	C
270	C	295	B	320	D	345	B	370	D	395	A	420	B	445	C	470	B	495	B
271	A	296	D	321	B	346	A	371	B	396	A	421	A	446	A	471	A	496	C
272	A	297	B	322	A	347	D	372	A	397	D	422	C	447	C	472	D	497	D
273	C	298	B	323	C	348	A	373	B	398	B	423	D	448	B	473	B	498	B
274	B	299	D	324	D	349	D	374	C	399	A	424	A	449	A	474	A	499	A
275	B	300	A	325	C	350	D	375	D	400	C	425	B	450	B	475	A	500	C